FROM
REJECTED STONE
— TO —
CORNERSTONE

A History of the Other Protestant Denominations
in Australian Military Chaplaincy

ROBERT VINCENT SMITH

SCD Press

From Rejected Stone to Cornerstone:
A History of the Other Protestant Denominations
in Australian Military Chaplaincy
Robert Vincent Smith

Occasional Series as No 3

© SCD Press 2024

SCD Press
PO Box 6110
Norwest NSW 2153
Australia
scdpress@scd.edu.au

ISBN-13: 978-1-925730-55-5 (Paperback)
ISBN-13: 978-1-925730-56-2 (E-book)

Cover design and typesetting by Lankshear Design.

This book is dedicated to
the chaplains of the Other Protestant
Denominations, whose story
I have tried to tell.

CONTENTS

Prologue 3
1. The Colonial and Federation Era: The Unwanted Stone................ 5
2. World War I: The Freshly Hewn Stone................................. 27
3. World War I: The Weathering of the Stone 47
4. Between the Wars: The Stone in Storage 83
5. World War II: The Reshaped Stone.................................... 107
6. World War II: The Re-weathering of the Stone 129
7. The Cold War: The Stone Stands Fast 159
8. Post-Vietnam: The Emerging Stone 183
9. The War on Terror: The Cornerstone.................................. 203
Conclusion 225
Abbreviations... 233
Bibliography.. 235
Index 253

PROLOGUE

CHAPLAINCY IN THE AUSTRALIAN armed forces officially began in 1912 with the appointment of chaplains from the Anglican, Presbyterian, Methodist and Roman Catholic Churches[1] to the Royal Australian Navy. A year later four Chaplains General for the Army were appointed from those same denominations. Prior to this, chaplaincy to the British Army units that garrisoned the Australian colonies from 1788 to 1870 was provided by colonial chaplains, and later by local civilian clergy. Parish clergy from the major churches continued to fill this role for the locally raised colonial forces that replaced the British units up to the time of Federation.

The influence of the major churches in the development of Australian military chaplaincy[2] was profound and dominated for decades. It compelled the Navy, which wanted a 'trans-denominational' form of chaplaincy, to adopt one based on the declared denominational affiliations of naval personnel. This virtually guaranteed Anglican dominance of naval chaplaincy and imposed the sectarianism of the day upon that service.

A year later the Army underwent a similar process and adopted a chaplaincy system that allowed for equal numbers of chaplains from the major churches. However, unlike the Navy, the Army permitted the appointment of chaplains from other recognised denominations, giving birth to what became the Other Protestant Denominations (OPD) in Australian military history. The inherent problem, though, was that

---

1 Hereafter referred to as 'the major churches'.
2 *Military* in this work is used in its broadest sense as 'pertaining to ... armed forces, affairs of war' (cf. Macquarie Dictionary).

such appointments were to be made on an 'As required' basis, which, given the overwhelming dominance of the major churches, meant that participation by the Other Protestant Denominations would be negligible and would not address the fundamental problem of shaping a denominationally based system to fit what was primarily a trans-denominational need.

Alluding to the Biblical references concerning the 'stone that was rejected by the builders' becoming the 'cornerstone',[3] this book records the history of the Other Protestant Denominations and its later iterations during seven major eras of Australian military history: the Colonial/Federation era, World War I, inter-war, World War II, Cold War, Post-Vietnam, and the two decades of high operational demand that began with the intervention in Timor-Leste and ended with the withdrawal from Afghanistan. It reveals how a numerically insignificant group, confronted by the toxic sectarianism that determined the original composition of the new nation's military chaplaincy, emerged from a marginalised position to play an increasingly influential and crucial role in its leadership. It encompasses a period of 230 years: from Richard Johnson's appointment as chaplain to the First Fleet and the Anglican dominance of military chaplaincy in Australia, to 2017 when the Associated Protestant Churches (successor to the Other Protestant Denominations) replaced the Anglican Church in that capacity.

The book's importance resides in both its historical value in filling in an important but largely untold part of the story of Australian military chaplaincy, and its contemporary relevance to current Defence Force chaplaincy confronted by increasing cultural and religious diversity. Furthermore, it articulates the historical process whereby the trans-denominational chaplaincy originally envisaged by the Royal Australian Navy – despite the opposition of the major churches – finally emerged in the *generic* form of chaplaincy characteristic of the OPD group. In this respect it has played a crucial part in enabling Australian Defence Force chaplaincy to meet the challenges of an increasingly diverse, pluralistic and rapidly changing social context.

---

3   Psalm 118:22, Matthew 21:42.

CHAPTER ONE

The Colonial and Federation Era: The Unwanted Stone

ON 18 JANUARY 1788 THE First Fleet of eleven vessels, commanded by Captain Arthur Phillip RN and carrying around fourteen hundred people, arrived at Botany Bay. Seven days later, having found Botany Bay unsuitable due to the anchorage being exposed to the prevailing south-easterly winds, the poor quality of the soil, and the inadequacy of the water supply,[1] Phillip sailed a few kilometres further north to Port Jackson and anchored at a place he named Sydney Cove. The following day, 26 January, accompanied by officers and men he went on shore, raised the Union Flag, and proclaimed British sovereignty over the eastern seaboard of Australia.[2] The formal proclamation of the colony of New South Wales took place on 7 February 1788 with Judge-Advocate David Collins performing the ceremony and Captain Phillip assuming the Office of Governor.[3] It was the culmination of a hazardous eight-month voyage by the fleet that had left Portsmouth on 13 May 1787.[4]

The new arrivals included an Anglican chaplain, Rev Richard

---

1 David Collins, *An Account of the English Colony in New South Wales*, Volume 1 (Sydney: A.H. & A.W. Reed, 1975), 3.
2 Arthur Phillip, 'Despatch to Lord Sydney, 15 May 1788; *Historical Records of NSW* Volume 1, Part 2 (Sydney: Charles Potter, Government Printer, 1892), 122.
3 Collins, *An Account of the English Colony*, 6.
4 Arthur Phillip, 'Despatch to Secretary Stephens, 12 May 1787; *HRNSW* Volume 1, Part 2, 103.

Johnson, who on 3 February, under a large tree in the vicinity of what is now Macquarie Place, conducted the first act of Christian worship on the east coast of Australia. The congregation included Governor Phillip, the Naval and Marine officers, their wives, and members of the Marine contingent.[5] Johnson, though, was not a military chaplain in the strict sense of the word. His commission from King George III appointed him 'Chaplain to the settlement within our territory called New South Wales', and even though he was to 'observe and follow such orders and directions as he should receive from time to time from our Governor ... according to the rules and discipline of war',[6] he was essentially a civilian and as such ministered to all members of the colony: military and convicts alike.

Johnson owed his post to a group of evangelical Anglicans known as the Eclectic Society,[7] who were concerned with prison reform and missionary work. One of their members was William Wilberforce, who introduced him to the Societies for the Propagation of the Gospel[8] and for Promoting Christian Knowledge,[9] which provided him with a large number of religious books and tracts to propagate the Christian message among the troops and convicts. His ministry was also influenced by Governor Phillip who, having listened to his sermons, urged him 'to begin with moral subjects' because, as chaplain to the settlement, he was required to be the guardian of public morality, which Phillip believed was his primary responsibility.[10] Balancing this with his duties as an Anglican clergyman under the general jurisdiction of the Bishop of London, and the evangelistic expectations of the Eclectic Society, whose protégé he was, caused Johnson considerable angst, and he never quite felt that he had succeeded. Phillip's 'rational approach to life and indifference to religious fervour' was unsympathetic to the

---

5   Kenneth Cable, 'Johnson, Richard (1753–1827),' *Australian Dictionary of Biography*, Volume 2 (Melbourne: MUP, 1967), 17.
6   George III, 'Commission to First Chaplain, 24 October 1786,' *HRNSW*, Volume 1, Part 2, 27.
7   A group of Anglican clergymen and laymen, established in 1783 as a discussion group, and instrumental in founding the Church Missionary Society in 1799.
8   Founded under royal charter in June 1701 as the official overseas missionary body of the Church of England.
9   Founded in 1698 to increase awareness of the Christian faith in England and across the world.
10 Cable, 'Johnson, Richard (1753–1827),' 17.

evangelicalism of Johnson,[11] whose letters home show how despondent he often felt.[12]

In this respect, Johnson foreshadowed the experience of later generations of military chaplains who found themselves 'servants of two masters', trying to fulfil their obligations as ministers of the churches that ordained them, and the state that paid them. Rowan Strong comments on this in relation to chaplains in the Royal Australian Navy.[13]

For twelve years Johnson faithfully 'performed the rites of passage, kept watch on public morality and conducted the statutory services at the compulsory church parades'.[14] He conducted services in the open air until a wattle and daub building was opened in 1793 – the precursor of St Philip's Church in what is now York Street, Sydney.[15] Hans Moll remarks that even though troops and convicts were obliged to attend Divine Service, Johnson only really had control over the convicts. In 1792 he complained to Governor Phillip: 'We are wholly exposed to the weather ... on this account, sir, it cannot be wondered that persons, whether of higher or lower rank, come so seldom and so reluctantly to public worship'.[16]

It was six years before his first assistant, Samuel Marsden, arrived.[17] Then, in 1800 when Johnson's failing health forced him to return to England, Marsden remained as the only Anglican clergyman on the mainland, eventually being appointed senior chaplain to the colony in 1810.[18] Marsden was also an Evangelical, but was a much more public figure than Johnson. He resolutely defended his rights, standing up to governors when he judged them to be interfering with his duties.[19] Over the next few years, he was joined by other Anglican chaplains including

---

11 B.H. Fletcher, 'Phillip, Arthur (1738–1814)', *ADB* Volume 2 (Melbourne: MUP, 1967), 327.
12 Ian Breward, *A History of the Australian Churches* (Sydney: Allen & Unwin, 1993), 13-14.
13 Rowan Strong, *Chaplains in the Royal Australian Navy, 1912 to the Vietnam War* (Sydney: UNSW Press, 2012), 20.
14 Stephen Judd and Kenneth Cable, *Sydney Anglicans* (Sydney: Anglican Information Office, 1987), 2.
15 Collins, *An Account of the English Colony*, 251; Allan Yuill, *Clerical Pioneers in New South Wales* (Sydney: Anglican Information Office, 1988), 3.
16 Hans Moll, *Religion in Australia* (Melbourne: Thomas Nelson, 1971), 9; Collins, *An Account of the English Colony*, 108.
17 Collins, *An Account of the English Colony*, 297.
18 A.T. Yarwood, 'Marsden, Samuel (1765–1838)', *ADB*, Volume 2 (Melbourne: MUP, 1967), 207.
19 Breward, *A History of the Australian Churches*, 14.

William Cowper and Robert Cartwright,[20] and together they did what they could to minister to free settlers, convicts and troops.

This pattern continued until the Church Acts in New South Wales and Tasmania in 1836 and 1837 gave Anglican, Roman Catholic, Presbyterian and Methodist churches equal access to government funding.[21] The designation 'colonial chaplain' then came to refer to clergy attached to the convict establishment,[22] and the resultant non-privileged status of the Church of England opened the way for those other denominations to become partners in the provision of military chaplaincy. It was a partnership they were to dominate until the latter part of the twentieth century.

Of the 24 British infantry regiments that served in Australia following the departure of the Marines,[23] only the New South Wales Corps arrived with a chaplain officially part of its strength. His name was James Bain, an Anglican clergyman,[24] and he arrived in Australia with one of the regiment's advance parties aboard HMS *Gorgon* on 21 September 1791.[25] The main body arrived with Major Grose on 14 February 1792.[26]

Bain's early arrival left him with limited military duties, and consequently Governor Phillip made use of his services among the wider community. In the *Returns of Troops* dated 22 November[27] and 15 December 1791,[28] he was recorded as being at Parramatta.[29] When Grose arrived with the main body of troops Bain was on Norfolk Island,

---

20 Andrew Houison, *A Short History of St Philip's Church Sydney* (Sydney: St. Philip's Vestry, 1910), 9-10.
21 Church Act 1837, in *Tegg's Pocket Almanac and Remembrancer* (Sydney: James Tegg, 1837), 101-102.
22 Michael Gladwin, *Captains of the Soul: A History of Australian Army Chaplains* (Sydney: Big Sky, 2013), 12.
23 Gladwin, *Captains of the Soul*, 13.
24 Sir George Yonge, 'Communiqué to Major Grose, 8 June 1789', *HRNSW*, Volume 1, Part 2, 250.
25 Philip King, 'Despatch to Under Secretary Nepean, 25 October 1791', *HRNSW*, Volume 1, Part 2, 529; Collins, *An Account of the English Colony*, 149.
26 Arthur Phillip, 'Despatch to Rt Hon Henry Dundas, 17 March 1792', *Historical Records of Australia*, Series 1, Volume 1 (Sydney: Library Committee of the Commonwealth Parliament, 1914), 336.
27 Arthur Phillip, 'Despatch to Lord Grenville, 22 November 1791', *HRA*, Series 1, Volume 1, 315.
28 Despatch to Under Secretary Nepean, 15 December 1791, in *HRA*, Series 1, Volume 1, 321.
29 Collins, *An Account of the English Colony*, 160.

sent there by Phillip[30] to act as a magistrate. He eventually returned to Sydney in February 1794.[31] Deteriorating health caused him to request a posting elsewhere and he returned to England, leaving Sydney on 17 December 1794.[32] No chaplain was sent to replace him.

The reason for this was that in 1796 the British Army abolished regimental chaplains, preferring to appoint them to higher formations.[33] Prior to this, regimental chaplaincies had been private arrangements between clergymen and regimental colonels. Gladwin comments: 'Like presentations to parochial livings in the Church of England, [they] were a form of property that could be bought or sold by an entrepreneurial colonel or, for that matter, an avaricious clergyman'.[34] Sir John Smyth's history of British Army chaplaincy mentions that such appointments were, prior to 1796, the 'perquisite of the Colonel'.[35]

So unsatisfactory was the system that in 1796, when the first Chaplain General, John Gamble, was appointed, he found that all three hundred and forty chaplains in his establishment were listed as being 'on leave'.[36] This probably came as no surprise in that when he was first appointed as a chaplain in 1793 he believed himself to be the only commissioned chaplain who was actually on full-time army service.[37] Gamble's advancement coincided with the Royal Warrant that established the Army Chaplains Department and decreed that chaplains should be appointed 'in the proportion of one to each brigade, or to every three of four regiments'.[38]

The New South Wales Corps garrisoned the colony until the arrival of Lieutenant Colonel Lachlan Macquarie with the 73rd Regiment (Royal Highlanders), on 28 December 1809. On 1 January 1810 he

---

30 Collins, *An Account of the English Colony*, 162.
31 Vivienne Parsons, 'Bain, James (1789–1794),' *ADB* Volume 1 (Melbourne: MUP, 1966), 49.
32 William Patterson, 'Despatch to Rt Hon Henry Dundas, 21 March 1795,' *HRA*, Series 1, Volume 1, 489.
33 A formation is a military disposition of troops.
34 Gladwin, *Captains of the Soul*, 4.
35 John Smyth, *In This Sign Conquer* (London: Mowbray, 1968), 26.
36 Roy Burley, 'The Age of Negligence? British Army Chaplaincy 1796–1844,' (Master's thesis, University of Birmingham, 2013), 39.
37 Smyth, *In This Sign Conquer*, 30.
38 Burley, 'The Age of Negligence?' 41.

officially became Governor of New South Wales.[39] Macquarie soon realised there was a need for more clergymen in the colony, and wrote to Earl Liverpool, Secretary of State for War and Colonies, asking for at least four more chaplains.[40] One result was the arrival in July 1814 of Benjamin Vale, who expected to take up a position as assistant chaplain to the recently arrived 46th Regiment (South Devonshire). However, Macquarie had not been informed of this and instead appointed him assistant to Rev William Cowper at St. Philip's Church, Sydney.[41] Vale later fell foul of Macquarie and was disciplined but defended his action to the point where Macquarie had him tried and convicted of 'conduct highly derogatory to his Sacred Character'. He escaped with a private reprimand and soon after returned to England.[42]

Apart from James Bain, colonial chaplains and civilian clergy provided chaplaincy to the British regiments that garrisoned Australia until 1870. The King's instructions to Governor Macquarie included the requirement that a high priority should be given to the observance of religion:

> It is our Royal will and pleasure that you, by all proper methods, enforce a due observance of religion and good order among the inhabitants of the said settlement, and that you do take particular care that all possible attention be paid to the due celebration of Public Worship.[43]

The degree to which governors fulfilled the spirit of this policy varied. Johnson grew tired of waiting for Governor Phillip to provide a church building and erected Australia's first house of worship from his own funds. Only after it was mysteriously burnt down a few years later did the Governor release funds to replace it.[44] Nevertheless, the Governors did from time to time issue orders that convicts, troops and

---

39 'Arrival of Lachlan Macquarie in Sydney, 28 December 1809,' *HRA*, Series 1, Volume 7, 181.
40 Lachlan Macquarie, 'Despatch to Earl Liverpool, 27 October 1810,' *HRA*, Series 1, Volume 7, 346.
41 Lachlan Macquarie, 'Despatch to Earl Bathurst, 4 December 1815,' *HRA*, Series 1, Volume 8, 299-300.
42 Lachlan Macquarie, 'Despatch to Earl Bathurst, 8 March 1816,' *HRA*, Series 1, Volume 9, 43-49.
43 George III, 'Governor Macquarie's Instructions, 9 May 1809,' *HRA*, Series 1, Volume 7, 192.
44 Yuill, *Clerical Pioneers in New South Wales*. 3.

free settlers should attend church services. Despite Marsden's antipathy and the ever-widening rift between them,[45] Macquarie's October 1810 despatch to Earl Liverpool proudly reported that during his short time in the colony there had been 'a very apparent Change for the better in the Religious Tendency and Morals of All the different Classes of the Community'.[46]

The End of the Anglican Monopoly

By 1829 the pattern of chaplaincy that Johnson had established was well embedded. Sydney then had a population of ten thousand and was served by two churches which had special sections allotted to the military.[47] There were also Methodist, Roman Catholic, and Presbyterian clergy present in the colony, but the extent to which they ministered to the garrison troops is difficult to assess. The first non-Anglican clergyman in Australia was the Wesleyan missionary Samuel Leigh, who arrived unofficially in August 1815. His arrival 'was welcomed by Marsden, who was himself an evangelical Anglican[48] but not by Macquarie, who at that stage had a very low opinion of Methodists'.[49] Macquarie's views are clear in his despatch to Lord Bathurst:

> But though Mr Leigh's Conduct has been hitherto very correct here, still I should strongly recommend that no Persons of his Description Should in future be permitted to Come over to this Colony. We require regular and pious clergymen of the Church of England, and not Sectaries.[50]

Eventually, Leigh was able to defuse Macquarie's initial hostility regarding his 'sectarian' status and the Governor, assured that Leigh was not

---

45 N.D. McLachlan, 'Macquarie, Lachlan (1762–1824),' *ADB* Volume 2 (Melbourne: MUP, 1967), 191-192.
46 Lachlan Macquarie, 'Despatch to Earl Liverpool, 27 October 1810,' *HRA*, Series 1, Volume 7, 346.
47 Judd and Cable, *Sydney Anglicans*, 11-12.
48 The Methodists separated from the Church of England in 1795.
49 Breward, *A History of the Australian Churches*, 16.
50 Lachlan Macquarie, 'Despatch to Earl Bathurst, 18 March 1816,' *HRA*, Series 1, Volume 9, 59.

about to challenge the authority of the Crown and the Church of England, gave full support for his ministry.[51]

Macquarie was far more incensed by the unauthorised arrival of Father Jeremiah O'Flynn (Roman Catholic) in November 1817. He ordered O'Flynn to return in the same ship, but the priest went into hiding and was subsequently deported in 1818.[52] He was followed in 1820 by Fathers John Therry and Philip Conolly, who were the first Roman Catholic priests to serve officially in Australia. Ian Breward notes that the resident Protestant clergy did not make it easy for them, and 'they had to contend with Protestant harassment over the right to conduct marriages and burials or to visit members of their flock in Protestant-run hospitals and orphanages'.[53]

The first Presbyterian minister was Rev John Dunmore Lang who disembarked in 1823. 'He was young, energetic, talented and disputatious, and took the authorities aback by demanding equal status with the Church of England'.[54] These pioneer clergy were followed by a succession of others, and through them the major churches became established in Australia. It is likely that their services were made available to soldiers who belonged to those denominations because the Anglican monopoly in British Army chaplaincy began to diminish soon after Lang's arrival. The British Army officially recognised the Presbyterian Church in 1827 and made it a separate branch of the Army Chaplains Department, and in 1836 also included Roman Catholic chaplains.[55]

Despite the sectarianism evident in the attitude of many Anglicans towards non-conformist Protestants[56] and the disdain they all felt towards Roman Catholics, it was Sir Richard Bourke's Church Acts of 1836 and 1837 that ensured that the Church of England would not become the 'established' church in Australia. Bourke, who was Governor of New South Wales from 1831 to 1837, was a liberal Anglican who

---

51 Elizabeth de Reland, 'Holiness and Hard Work: A History of Parramatta Mission, 1815–2015,' (PhD thesis, Charles Sturt University, 2018), 56.
52 Breward, *A History of the Australian Churches*, 16.
53 Breward, *A History of the Australian Churches*, 16-17.
54 Breward, *A History of the Australian Churches*, 19; D. Baker. 'Lang, John Dunmore (1799–1878),' *ADB* Volume 2 (Melbourne: MUP, 1967), 76-83.
55 Smyth, *In This Sign Conquer*, 28.
56 Breward, *A History of the Australian Churches*, 16.

abhorred sectarian intolerance, and believed that any attempt to establish a dominant colonial church was doomed to failure.[57] His Church Acts recognised the right of all recognised denominations to receive financial aid from the colonial administration should there be at least one hundred adults in a community declaring their desire to attend a particular church or chapel.[58] This effectively put each of the major churches on an equal footing. Eighty years later it was a significant factor in shaping the composition of Australian military chaplaincy.

The Other Protestant Denominations – at that time mainly Baptists and Congregationalists, who 'were used to self-help' – tended not to avail themselves of government support as a matter of principle. It was commonly felt among them that to accept money from the government would compromise their independence and reduce the ministry to just another way of making a living.[59] This, and their relative insignificance in terms of size, caused them to be overlooked in ecclesiastical negotiations with government agencies, including those that later produced the official military chaplaincy organisations.[60]

Consequently, the absence of dedicated regimental chaplains to the British regiments that garrisoned Australia – apart from the New South Wales Corps[61] – and the relative insignificance of the smaller Protestant denominations meant that pastoral and spiritual care of the troops was provided on an *ad hoc* basis by civilian clergy of the major churches, in whose parishes the troops were stationed. These were 'unofficial appointments and were not formally gazetted on Army lists'.[62] Some acted as garrison chaplains and were paid small allowances. Tom Johnstone reports that throughout the colonial era in Australia Catholic priests 'applied for and received five shillings a week from the War Office for ministering to the regiments of the colonial garrisons' and that 'it formed the main part of their income'.[63]

---

57 King, H. *'Bourke, Sir Richard,* (1777–1855),' *ADB* Volume 1 (Melbourne: MUP, 1966), 131.
58 Church Act 1837, *in Tegg's Pocket Almanac,* 101-102.
59 Breward, *A History of the Australian Churches,* 38.
60 They were followed later by the Churches of Christ and the Salvation Army.
61 A search of the regiments' records held in the Mitchell Library revealed no names of chaplains.
62 Gladwin, *Captains of the Soul,* 13.
63 Tom Johnstone, *The Cross of Anzac, Australian Catholic Service Chaplains* (Brisbane: Church Archivists' Press), 1.

The First Colonial Forces

The Crimean War of 1853–1856 was a catalyst for defence preparedness and the raising of local forces,[64] but colonial administrations did not seriously begin to recruit local militia units until after the New Zealand War of 1863–1864.[65] It was then that the first chaplain was appointed to a military force that was solely Australian. Arthur Bottrell's research discovered that in 1862 Rev Rowland Davies, Anglican Archdeacon of Hobart, became chaplain to Tasmania's Volunteer Force. Later that year he was joined by Rev John Storie, a Presbyterian.[66] Gladwin says that in following years they were joined by Roman Catholic, Methodist and Congregational chaplains.[67] The reference to the latter is significant in that it would mark the first entry of an OPD chaplain into Australian military chaplaincy.[68]

In 1869 the colony of Victoria informed the British government that it could no longer contribute to the support of British troops unless there was an increase in artillery and guarantees of troop deployments in times of crisis. The following year the Secretary of State for the Colonies responded by advising that at the end of September the last British troops would be withdrawn, and the Australian colonies would become responsible for their own locally raised defence forces.[69] New South Wales, in 1871, was the first to raise a permanent paid defence force in the form of a battery of artillery and two companies of infantry. This was in addition to twenty-eight companies of unpaid volunteer rifles and nine batteries of volunteer artillery. Victoria followed with a small permanent unit in addition to its volunteers, and also purchased the ironclad monitor *Cerberus*.[70]

In 1876, the governments of New South Wales, Victoria, Queensland and South Australia asked the British to provide an imperial officer to

---

64 Jeffrey Grey, *A Military History of Australia* (Melbourne: CUP, 2008), 22.
65 Gladwin, *Captains of the Soul*, 13.
66 Arthur Bottrell, 'Australia's first two commissioned chaplains,' *Intercom*, no 29 (December 1983), 13-15.
67 Gladwin, *Captains of the Soul*, 13.
68 A search of records in the State Library and Archives Service of Tasmania was unable to identify him.
69 Grey, *A Military History of Australia*, 23.
70 Grey, *A Military History of Australia*, 42-43.

give 'good professional advice' on the state of their defences. The British responded by deploying Major General Sir William Jervois and Lieutenant Colonel Peter Scratchley to Australia to advise on a coordinated plan of defence.[71] Their recommendations subsequently became the basis of Australia's colonial defence until the Commonwealth of Australia assumed control in 1901.

The first chaplains to the volunteer forces in Victoria were also appointed in 1876 and by 1900 had grown to forty-seven: thirty-four Anglican, seven Presbyterian, three Roman Catholic, two Methodist, and one Congregational.[72] Again, the reference to a chaplain from the Congregational Church is noteworthy.[73] In New South Wales an establishment of one Senior Chaplain and eleven chaplains was promulgated in 1895: seven Anglicans, two Presbyterians, one Methodist and one Roman Catholic.[74]

The first Australians to serve overseas were fourteen hundred and seventy-five volunteers who, in 1863, went to fight in the Maori War.[75] No chaplain went with them.[76] Then, in 1885, the New South Wales Government sent a contingent of colonial troops to the Sudan in support of Britain's campaign against the radical Islamic Mahdist movement. The Anglican, Roman Catholic, and Wesleyan churches all offered chaplains for this force, and the government, bowing to public pressure, eventually agreed to send one Anglican and one Roman Catholic.[77] Rev Herbert John Rose[78] was the Anglican chaplain and Father Charles Collingridge[79] the Roman Catholic. Rose provided chaplaincy support to men of all the Protestant denominations and Collingridge to the Roman Catholics. They did, however, cooperate in organising recreational activities for the troops – an encouraging

---

71 Grey, *A Military History of Australia*, 44.
72 Johnstone, *The Cross of Anzac*, 3.
73 A search of the archives of the Public Record Office of Victoria failed to identify him.
74 Johnstone, *The Cross of Anzac*, 3.
75 Peter Firkins, *The Australians in Nine Wars, Waikato to Long Tan* (Sydney: Pan, 1982), 3.
76 Johnstone, *The Cross of Anzac*, 2.
77 Gladwin, *Captains of the Soul*, 14.
78 Malcolm Saunders, 'Rose, Herbert John, (1857–1930),' *ADB* Volume 11 (Melbourne: MUP, 1988), 449.
79 Johnstone, *The Cross of Anzac*, 3.

sign of the unofficial ecumenical cooperation that later characterised Australian military chaplains.[80]

The final event to shape Australia's colonial military history was the Boer War which began in 1899. Australia offered the British government a force of two thousand five hundred men. In addition to feelings of solidarity with the mother country there was also widespread support for the Uitlanders. These were British subjects living in the Transvaal who were denied citizenship by the Boers and who, in many ways, were like Australian colonists.[81] Despite British concerns about the reliability of colonial troops, by the war's end more than sixteen thousand Australians had served in South Africa, earning an enviable reputation.[82] Jeffrey Grey attributes this mostly to the first two Australian contingents, 'drawn largely from the part-time militias of the colonies'.[83] Among them was Arthur Forbes, a bugler who won the Distinguished Conduct Medal, and later served as an OPD chaplain in both World Wars.[84]

Peter Firkins notes that the Boer War 'initiated numerous aspects of military practice which are regarded as fundamental today'. One was the attachment of chaplains to individual units rather than to larger formations, a practice not followed by the British Army (except for Scottish and Irish regiments which had, *de facto*, their own chaplains) until after the Boer War.[85] The eighteen Australian chaplains who served in that conflict were all from the major churches.[86] Gladwin says of them:

> Their work during the war set the template for later chaplaincy on overseas service: providing church parades and voluntary worship services; ministering to the sick and wounded ... administering last rites, burying the dead and for their internment [sic] in war cemeteries.[87]

---

80 Gladwin, *Captains of the Soul*, 14.
81 Thomas Pakenham, *The Boer War* (London: Abacus, 1992), 249.
82 Firkins, *The Australians in Nine Wars,* 9-16.
83 Grey, A Military History of Australia, 63.
84 Dennis Nutt, 'Military Chaplains: For Service of our Soldiers', *AACJ* 27, 2016, 19-31.
85 Firkins, *The Australians in Nine Wars,* 16.
86 F.L. Murray, (ed.) *Official Records of Australian Military Contingents to the War in South Africa, 1899–1902* (Melbourne: Department of Defence, 1911), 580-607.
87 Gladwin, *Captains of the Soul*, 16.

Naval Chaplaincy in the Colonial Era

As with the army, chaplaincy in the Royal Australian Navy was modelled on its British parent. From the very beginning of Australia's colonial era the Royal Navy was a significant influence. The first four Governors of New South Wales were captains in the Royal Navy, as were several other colonial governors. They all eventually became admirals under the prevailing system whereby if an officer held post rank and lived long enough, he would reach Flag rank.[88]

One naval chaplain who made a significant contribution to the early religious life of two Australian colonies was Robert Knopwood, chaplain of HMS *Resolution*. He joined David Collins' expedition to Port Phillip and conducted the first Christian service there in 1803, returning with Collins to Van Diemen's Land in 1804 where he officiated at the first service in Hobart Town.[89] Apart from him there is very little reference to Royal Navy chaplains in Australian colonial history.

Strong observes that there were at least two Royal Navy chaplains aboard ships of the Australia Station. Charles Haslewood, chaplain of HMS *Orpheus*, lost his life when the vessel sank at the entrance to Manakau harbour in Auckland during the Maori War. He was probably the first chaplain to die on Australian service. The other was Samuel Payne, chaplain of HMS *Curacoa*, the flagship of the Australia Station between 1863 and 1867. He went on shore with men of the ship's company during the Waikato Campaign.[90]

Along with land forces, the Australian colonies began to develop small naval forces in the second half of the nineteenth century, especially from the 1880s when potential threats from France, Russia and Germany emerged. The Colonial Defence Act of 1865 saw Queensland, New South Wales, Victoria, Tasmania and South Australia set up their own tiny navies, manned by very small permanent forces, supported by a larger volunteer Naval Reserve.[91] Strong states that 'Chaplains also served in the colonial navies ... but these were all reservists, like most

---

88   Steve Pope, *Hornblower's Navy: Life at Sea in the Age of Nelson* (London: Orion, 1998), 61.
89   Strong, *Chaplains in the Royal Australian Navy*, 37.
90   Strong, *Chaplains in the Royal Australian Navy*, 38.
91   The Colonial Naval Defence Act 1865 allowed the colonial governments to own ships but under the Royal Navy's command.

of their compatriots'.[92] He also reports that though some colonial navy sailors saw action in various imperial wars, it does not appear that any colonial naval chaplain accompanied them.[93]

Chaplaincy to the Australian Army after Federation

Military chaplaincy during the colonial era provided the antecedents of the chaplaincy organisations in the three Australian armed services. Australia became a nation when the six former colonies united to become the Commonwealth of Australia on 1 January 1901. Two months later, on 1 March 1901, the naval and military forces of the former colonies were transferred to the Commonwealth, which was given the power 'to make laws with respect to the naval and military defence of the Commonwealth'.[94]

The forces the Commonwealth inherited were small and disorganised. The Army consisted of just over twenty-eight thousand men, of whom only fifteen hundred were permanent soldiers. The Navy had around two thousand, with two hundred and fifty on full-time duty 'manning an assorted collection of vessels in various stages of obsolescence'.[95]

The man appointed head of the new Australian Army on 26 December 1901 was Major General Sir Edward Hutton. He thought that Australia's best interests would be served by a national army of citizen soldiers, which he believed was 'the true form for an army for an Anglo-Saxon state to possess, a large standing army being an unnecessary and unwarranted expense'.[96] The Commonwealth Defence Act[97] gave him a statutory base on which to operate.[98] Under his leadership the Australian Army began to take shape. A general staff system was adopted and with it came the reorganising of the various States into more convenient Military Districts.[99]

---

92 Strong, *Chaplains in the Royal Australian Navy*, 38.
93 Strong, *Chaplains in the Royal Australian Navy*, 34.
94 Commonwealth of Australia Constitution, section 51(vi).
95 Grey, *A Military History of Australia*, 67.
96 Grey, *A Military History of Australia*, 68.
97 Commonwealth of Australia Defence Act, 1903.
98 'Defence of Australia: Memorandum', *CPP*, General Session 1910, Volume II, 83-104.
99 Johnstone, *The Cross of Anzac*, 12.

Following Hutton's concept of a 'national army of citizen soldiers', the Citizen Military Forces (CMF) – or Militia – became an integral part of the evolving Australian Army. This system produced some of the most famous of Australia's future military commanders of World War I and World War II – something that was often resented by the professional officer corps.[100] It also marked a difference between the emerging administration of Australian Army chaplaincy under the oversight of a conference of part-time Chaplains General, and that of its British Army parent which was led by one full-time Anglican Chaplain General.[101]

Chaplaincy, however, was a low priority in the reorganisation of the Australian Army, and no new chaplains were appointed to the pre-Federation list for some time.[102] The most critical assessment of this came through a retired British Chaplain General, John Cox Edgehill, at an Anglican Church Congress in Brighton (England) in 1901. He described Australian military chaplaincy as having suffered from lack of proper organisation and filled with anomalies. He reported that 'some chaplains were honorary; others were appointed to volunteer forces only, yet others to particular regiments or military forces generally; some were paid; and there was confusion about relative rank and dress regulations'.[103]

A conference of New South Wales chaplains in February 1902 responded to this by drafting suggested changes to the higher organisation of chaplaincy. It proposed a Senior Chaplain for each of the major churches. In consultation with his official denominational leader, he would recommend appointments and promotions and convene quarterly consultations of all military chaplains within the military district. Further, all chaplains, whether permanent, partially paid or of the Volunteer Forces, should be remunerated.[104]

Henry Mort, the only Chaplain 1st Class (Colonel) at the time, wrote to the Minister of Defence on 16 November 1904 submitting a draft for the organisation of a Chaplains Department based on these

---

100 The author often heard Regular Army officers dismissively refer to Army Reserve Generals as the *Rum Corps*.
101 The first non-Anglican Chaplain General of the British Army was a minister of the Church of Scotland and was appointed in 1986.
102 Johnstone, *The Cross of Anzac*, 12.
103 Gladwin, *Captains of the Soul*, 24-25.
104 Gladwin, *Captains of the Soul*, 25.

recommendations. It included the suggestion that there should be a Principal Chaplain for each of the major churches, and that 'the Senior Chaplain in Australia, of whatever denomination, be called the 'Principal Chaplain of the Commonwealth'.[105] This was similar to the structure of chaplaincy in the British Army, except that the British Chaplain General was required to be a minister of the Church of England. Nevertheless, even though the basis of an organized chaplaincy system was present in these initiatives,[106] Gladwin notes that the Minister for Defence, Robert Collins, 'simply crossed out the paragraph about Principal Chaplains and ... With a stroke of his pencil and the placement of the letter on file ... helped to delay the creation of a Chaplains' Department for almost another decade'.[107]

Being aware that the New South Wales chaplains had only proposed, but not voted on, the new leadership structure,[108] it is likely that the Minister's response may have been due to anxiety about potential sectarian squabbles. This was evident in a deputation from the Moderator of the Presbyterian Assembly of New South Wales in 1905, protesting that:

> It has become obnoxious to us that we have to communicate through the chaplains of other denominations. We prefer to make our nominations direct to the Defence Department without the intervention of any third party.[109]

Despite this the relationship of chaplaincy to the Military Board of Administration was still unclear. The situation of the Army Chaplains Department under the 1904 Regulations was 'somewhat ethereal' in that it was not listed among the corps, and no member of the Military Board of Administration had responsibility for it. It was not until the 1908 Regulations that chaplains became the responsibility of the Adjutant General.[110]

However, this 'ethereal' and disorganised situation that John Cox

---

105 Douglas Abbott, 'In This Sign Conquer: The Chaplains General of the Australian Army, 1913–1981'. Unpublished manuscript, 1995, 47.
106 Organisation of Chaplains Department Military, NAA B168, 1904/5368.
107 Gladwin, *Captains of the Soul*, 26.
108 Mort's letter had emphasised this: Abbott, *In This Sign Conquer*, 47.
109 Attachments relating to Military Chaplains, NAA B168, 1905/5614.
110 Abbott, *In This Sign Conquer*, 46.

Edgehill had described as 'full of anomalies' did have an OPD component in the person of Rev Alfred Metters (Baptist), who was appointed as a chaplain in the CMF in 1906, making him, apart from the two unknown Congregationalists of the colonial era, the first OPD chaplain. He served in South Australia and, following the formation of the Other Protestant Denominations as a recognised group within army chaplaincy, was appointed OPD Senior Chaplain for his military district. His reputation was such that when his two-year appointment ended his fellow OPD chaplains unanimously recommended his reappointment, 'as a mark of their appreciation of the services he had rendered'. On the outbreak of war, he offered himself for active service but was rejected on the grounds of being over the military age. Instead, he gave himself unstintingly to the welfare of soldiers at the Mitcham army camp and was renowned for having witnessed the departure of nearly every South Australian to the Front and being present to welcome every one of them who returned home.[111]

Eventually it was another factor that brought matters to a head. In 1909 the Australian Government invited Lord Kitchener – the most revered soldier in the British Empire – to inspect Australia's defences. On his advice a system of compulsory military training was established and began in 1911.[112] There followed a rapid expansion of the Army as more than ninety thousand young men reported for training,[113] raising concerns in the churches about how best to provide chaplaincy support for them. This led to the appointment of four more OPD chaplains: Rev Frederick Miles (Baptist) and Rev E Davies (Congregational) as senior chaplains for 3rd Military District,[114] George Walden (Churches of Christ) to the 4th Military District in September 1913,[115] and William Lamb (Baptist) to the 2nd Military District in May 1914. Six other Baptists were appointed in 1914 to serve 'the boys' attending periods of compulsory training.[116]

---

111 *Australian Baptist*, 31 July 1917, 3.
112 Grey, *A Military History of Australia*, 78.
113 'Defence of Australia: Memorandum', *CPP*, General Session 1910, Volume II, 83-104.
114 Their names appear on a letter written by the Assistant Adjutant General, dated 11 July 1911, Conference Proceedings, NAA A2023, A82/1/24.
115 Nutt, 'Military Chaplains', 20. Walden was the first chaplain from Churches of Christ.
116 *Australian Baptist*, 29 December 1914, 4.

On 5 March 1913, a most significant milestone was reached when representatives of the Roman Catholic Church met with the Adjutant General, Lieutenant Colonel Harry Chauvel, followed on 31 March and 5 April by representatives of the Anglican, Presbyterian and Methodist denominations. These meetings formulated the shape of the Australian Army Chaplains Department. There would be four Chaplains General – one each from the Anglican, Roman Catholic, Presbyterian and Methodist Churches, selected by the Australian head of each denomination. There would also be a Senior Chaplain appointed in each Military District for each of those denominations, whose responsibility would be to administer chaplains of his own denomination.[117] Once more, sectarianism determined that military chaplaincy would be focused denominationally rather than ecumenically.

Gladwin describes the agreements reached in the two sets of meetings as 'a kind of "magna carta" of the Australian Army Chaplains Department'.[118] The total establishment of one hundred and sixteen chaplains, however, only specified the appointment of chaplains from the major churches. It included four Chaplains General at Army Headquarters, and twenty-four Senior Chaplains – four to each of the six Military Districts, divided equally between those denominations. The remaining eighty-eight chaplains were to be distributed equally across twenty Brigades, except for the 1st and 23rd Brigade Areas which were to have two chaplains from each denomination.[119]

There was, however, another outcome of the meetings with the Adjutant General that had important implications for chaplaincy in the Australian Army. Section 112(b) included the stipulation that clergymen of "any recognised religious body" (to be designated "Other denominations") could be appointed chaplains on the authorised establishment on the recommendation of District Commandants and the Military Board'. Section 112(c) then included 'Other denominations' as the fifth category following the Anglican, Roman Catholic,

---

117 Appointment of Chaplains to Salvation Army and Other denominations, NAA A2023, A82/1/24.
118 Gladwin, *Captains of the Soul*, 28.
119 Chaplains Conference Proceedings: Revision of Regulations, NAA A2023, A82/1/24.

Presbyterian and Methodist denominations.[120] This decision marks the entry of what was soon to be known as the Other Protestant Denominations into Australian military chaplaincy.

The appointment of the four Chaplains General was promulgated on 20 December 1913.[121] As previously noted, this model was a clear departure from the practice of the British Army Chaplains Department from which the Australian Army Chaplains Department derived its ethos. It reflected the Australian abhorrence of the idea of an established church. Unfortunately, while maintaining a semblance of equality between the major churches, it was ill-suited to the way military organisations operate. Gladwin proffers that 'it would later prove to be a source of embarrassment due to its enshrining of intractable sectarian difference and its departure from the hierarchical leadership structure of the Army'.[122]

It took another seven decades before this anomaly was rectified by the replacement of the Conference of Chaplains General with a committee of three Principal Chaplains, representing the Anglican, Roman Catholic, and Protestant Churches, one of whom was, on a rotating basis, designated Principal Chaplain-Army, and became the official Head of Corps.[123] The creation of a Conference of Chaplains General, however, for all its shortcomings was inevitable, given the denominational sensitivities of the time. It maintained the principle of equal status between the major churches, while leaving the door open to the participation of other smaller denominations 'as necessary'.

The practice of appointing equal numbers of chaplains from the major churches changed in 1914 to one of proportional representation, based on religious affiliation as reported in the 1911 census.[124] The actual figures reported in that census were: Anglican 38.4%; Roman Catholic 22.4%; Methodist 12.3%; Presbyterian 12.5%; Baptist 2.17%; Congregational 1.66%; Churches of Christ 0.87%; Salvation Army

---

120 Chaplains Conference Proceedings: Revision of Regulations, NAA A2023, A82/1/24.
121 Commonwealth of Australia *Gazette*, 20 December 1913, Executive Minute 943, 3320.
122 Gladwin, *Captains of the Soul*, 29.
123 Gladwin, *Captains of the Soul*, 260.
124 Michael McKernan, *Australian Churches at War: Attitudes and Activities of the Major Churches 1914–1918* (Sydney: Catholic Theological Faculty and Australian War Memorial, 1980), 41.

0.6%, and Lutheran 1.62%.[125] It was a system that disadvantaged the Other Protestant Denominations because, as Abbott notes: 'the denominations showing larger percentages often carried a heavy nominal membership',[126] something Tom Frame describes as 'a source of acute embarrassment' to Anglican and Protestant church leaders.[127] Nevertheless, Australian Army chaplaincy and the place of the Other Protestant Denominations in it had taken a huge step forward.

Chaplaincy to the Royal Australian Navy

One of the factors that precipitated the organisation of chaplaincy in the Australian Army is that the Navy had already started to establish its Chaplaincy Branch. The Commonwealth Defence Act of 1904[128] gave Australian naval forces an overall commander, and a Board of Naval Administration soon followed. Strong points out that 'a navy founded from the RN had to have chaplains, and consequently it was only a short time after the inauguration of the RAN in 1911 that moves began towards establishing an Australian naval chaplaincy'.[129] Furthermore, reservist chaplains in the former colonial forces, like all other personnel, had already been enrolled in the Commonwealth Naval Forces after Federation.[130]

In 1911 an Australian Commonwealth Naval Board became the administrative authority for the new navy which was divided into Permanent Naval Forces with Citizen Naval Forces in reserve.[131] Moves towards establishing an official naval chaplaincy began almost immediately, culminating in an agreement between representatives of the Anglican, Presbyterian and Methodist Churches, and the Naval Director and Naval Secretary. It was agreed that chaplaincy appointments would be made by a committee representing those three denominations

---

125 Commonwealth of Australia Census, 3 April 1911.
126 Abbott, *In This Sign Conquer*, 51.
127 Tom Frame, *Losing My Religion – Unbelief in Australia* (Sydney: UNSWP, 2009), 54.
128 'Defence of Australia: Memorandum', *CPP* General Session 1910, Volume II, 83-104.
129 Strong, *Chaplains in the Royal Australian Navy*, 40.
130 Strong, *Chaplains in the Royal Australian Navy*, 35.
131 Designation of the Commonwealth Naval Forces as Royal Australian Navy, NAA MP1185/9, 559/201/574.

on the basis of census figures – three Anglican to one Presbyterian and one Methodist. A second meeting with representatives of the Roman Catholic Church agreed that 'for every four appointments three would be Protestant and one Roman Catholic'.[132]

This was another departure from the British practice. The Church of England had a virtual monopoly on chaplaincy in the Royal Navy during the eighteenth and nineteenth centuries. At the time of Australia's colonial foundation chaplains in the Royal Navy, though strictly warrant officers rather than commissioned officers, were recognised as gentlemen and usually messed with commissioned officers. From 1812 they were eligible to retire on half pay after eight years' service.[133] It was not until 1904 that Church of Scotland chaplains were appointed, serving on an honorary basis until World War I, and the first Roman Catholic and Wesleyan chaplains were only commissioned in 1918.[134] Church of England customs and religion dominated the culture of the Royal Navy long after the British Army opened its chaplaincy to non-Church of England clergy.[135]

Despite the agreements made with the major churches, as with Army chaplaincy, it was evident that the administration of the chaplaincy system did not comfortably fit the Navy culture. The Second Naval Member, Captain Chambers, whose portfolio included personnel, emphasised that the Navy wanted a 'trans-denominational' (ecumenical) form of chaplaincy. A ship's chaplain had to be available to all members of the ship's company. But this 'naval trans-denominationalism was at odds with the denominationalism of contemporary Christianity'.[136]

The eventual outcome of the agreements between the Navy and the major churches was something less than the generic form of chaplaincy the Navy wanted. Roman Catholic Chaplains would minister solely to Roman Catholics, while those of the other three Churches would act as

---

132 Conference on appointment of Roman Catholic Chaplains and Protestant to Royal Australian Navy in the Defence force, NAA A2023, A82/1/2.
133 Pope, *Hornblower's Navy*, 54.
134 Civilian Roman Catholic priests had been allowed to officiate since 1887; Strong, *Chaplains in the Royal Australian Navy*, 19-20.
135 Strong, *Chaplains in the Royal Australian Navy*, 20.
136 Strong, *Chaplains in the Royal Australian Navy*, 42.

'pan-Protestant' chaplains.[137] Inadequate though it was, it reflected the uneasy nature of inter-Church relationships at the time and a reluctance by the major churches to give ground to each other. It is therefore little wonder that several decades were to pass before chaplains of the Other Protestant Denominations would gain access to the Royal Australian Navy.

For military chaplaincy the Colonial and Federation eras had been a long and tortuous journey. Beginning with Richard Johnson and the early Anglican colonial chaplains, it passed on to the civilian clergy who ministered to the British garrisons and then to the 'ungazetted' volunteers of the later colonial years. It culminated in the chaplains from the major churches who served in the Boer War. By the outbreak of World War I Australia's two armed services did have functioning chaplaincy organisations which, despite their sectarian limitations, in the crucible of war served them well. The early Anglican monopoly of chaplaincy had given way to a multi-denominational system that gave equal status to the major churches, and that by 1913 had opened the door slightly for the Other Protestant Denominations to also be involved. The unwanted stone had at last been recognised.

---

137 Strong, *Chaplains in the Royal Australian Navy*, 49.

CHAPTER TWO

World War I: The Freshly Hewn Stone Response to the War by the Major Churches

THE DECLARATION OF WAR by the British government on 4 August 1914 was greeted in Australia with unparalleled enthusiasm.[1] Bill Gammage, describing the jubilation of cheering crowds gathering in the streets, and men at Labor Party Headquarters and Melbourne University singing 'Rule Britannia', called it 'the most complete and enthusiastic harmony in their history'.[2] *The Age*, referring to the crowds gathered outside its Melbourne office, reported: 'Excited Crowds in Street Waving Union Jack'.[3] The *Sydney Morning Herald* two days later proclaimed: 'For good or ill, we are engaged with the mother country in fighting for liberty and peace ... It is our baptism of fire ... and the discipline will help us to find ourselves'.[4]

The whole nation, including its churches, became consumed by patriotic jingoism. Michael McKernan, with the benefit of historical hindsight, reflected that: 'The extent of the coming tragedy caused no church leader to draw back, rather they acquiesced in the approach

---

1 'The First Away,' in *The RSL Book of World War 1*, edited by John Gatfield with Richard Landels, (Sydney: Harper Collins, 2015), 7.
2 Bill Gammage, *The Broken Years, Australian Soldiers in the Great War* (Melbourne: Penguin, 1982), 4.
3 *Melbourne Age,* 5 August 1914, 8.
4 *Sydney Morning Herald*, 6 August 1914, 6.

of war'.⁵ The editors of denominational newspapers supported them, describing Germany as 'an ambitious and aggressive power, ready to break her word to impose militarism in Europe'.⁶ The *Methodist*, four days after the declaration of war, declared: 'every consideration of national obligation and honour seems to demand that she [Britain] should oppose with all her naval and military might the dangerous and aggressive attitude of Germany'.⁷ There was widespread condemnation of Germany's violation of Belgian neutrality, which Britain was pledged to defend,⁸ and the *Catholic Church Press* drew attention to Home Secretary Sir Edward Grey's report to the House of Commons of 'the King of Belgium's appeal to Britain to safeguard Belgian integrity'.⁹

In Sydney, on Sunday 9 August, at a special service in St Andrew's Cathedral Archbishop John Charles Wright judged 'a disgraceful peace as being more dreadful than war'. Speaking of the justness of the British cause he said: 'We can, with a clear conscience, ask God to forgive our many national and personal sins, yet claim to be on the side of God'.¹⁰ Similarly, in a united service of intercession that same day, Rev Patrick Stephen of the Central Methodist Mission assured the twenty thousand people gathered in the Sydney Domain that Britain had been drawn to declare war in order to help countries being attacked 'by an arrogant nation'.¹¹ The Dean of Sydney followed this 'with an impassioned address' and appealed to Australians 'not to waver in their patriotism'.¹² Eleven months later, despite the appalling casualties suffered, Archbishop Wright continued to preach 'the solemn call of duty as the voice of God'.¹³

For most churchmen there was an added dimension to their patriotic support. They believed the crucible of war would bring the nation

---

5 Michael McKernan, *Australian Churches at War: Attitudes and Activities of the Major Churches 1914–1918* (Sydney: Catholic Theological Faculty and Australian War Memorial, 1980), 24.
6 McKernan, *Australian Churches at War*, 34.
7 *The Methodist*, 8 August 1914, 7.
8 Article 7 of the Treaty of London (1839): 'Belgium ... shall form an Independent and perpetually Neutral State. It shall be bound to observe such Neutrality towards all other States'.
9 *Catholic Church Press*, 6 August 1914, 19.
10 *Daily Telegraph*, 10 August 1914, 9.
11 Wrongly named in the *Daily Telegraph* as Rev P.J. Stephens of the Central Methodist Union.
12 *Daily Telegraph*, 10 August 1914, 9.
13 *Australian Church Record*, 2 July 1915, 2.

back to the true path of righteousness and devotion to God. It was a commonly held view in churches at that time, drawing on passages from the Old Testament warning the nation of Israel that disobedience to God's commands would result in disease, disasters, and destruction.[14] McKernan claims that at that time churchmen by and large 'accepted war as one of the methods God used to chastise the people and alert them to the true path of devotion and duty'.[15] Naïve though this may be, in 1914 the full horror of the tragedy to come had not yet challenged those simplistic assumptions.

The Roman Catholic Church responded in much the same way as the Protestant Churches. Sydney Archbishop Michael Kelly spoke for many when he urged people to see the war 'as a chastisement from God for abandoning the true principles of righteousness and religion'.[16] McKernan, however, posits that though 'Catholic preachers gave the appearance of outward harmony with their Protestant counterparts in their views about the war, there were subtle differences which contained the seeds of future dispute'.[17] These differences reflected the Irishness of the Roman Catholic Church in Australia, and its sense of being a somewhat despised minority in a largely Protestant nation. According to Jeff Kildea World War I presented both a challenge and an opportunity to Australian Catholics. The challenge was to remain loyal to Ireland, especially after the British suppression of the 1916 Easter uprising. The opportunity was that by supporting Australia's involvement they might be accepted as equal members of the Australian community. Sadly, that did not happen. Sectarian bitterness was raised to levels previously unseen following the Easter uprising and the conscription campaigns.[18] But at the outset Catholic support for the war was widespread. Even Melbourne Archbishop Daniel Mannix, who campaigned strongly against conscription in the referenda of 1916 and 1917, and who by 1918 'had become the most reviled figure in Australian history,'[19]

---

14 See Leviticus 26:25-33, Isaiah 1:20, Jeremiah 5:17.
15 McKernan, *Australian Churches at War*, 24-25.
16 *Freeman's Journal*, 6 August 1914, 26.
17 McKernan, *Australian Churches at War*, 30.
18 Jeff Kildea, 'What Price Loyalty? Australian Catholics in the First World War,' *ACR*, 96 (2019), 43.
19 James Griffin, 'Mannix, Daniel (1864-1963),' *ADB* Volume 10 (Melbourne: MUP, 1986), 400.

initially supported involvement,[20] and 'hoped that the efforts of the allies would be crowned with success and the result would bring honour to the British Empire'.[21]

Across the nation churches organised services of intercession to pray that God's righteous purposes might be fulfilled. In Sydney, Protestant clergymen arranged daily united services of intercession at each of the major city churches on alternate days.[22] Patriotism and piety joined hands and pulpits were often draped with the Union Jack.[23] Common to many of those services was the theme of national spiritual renewal emerging from the morass of war, as in the sermon preached by the Anglican Dean of Perth, Rev Henry Mercer, who told his congregation: 'It was better to ride a warhorse than a racehorse'.[24]

Shakespeare's 'God of battles'[25] was invoked in a manifesto produced by Brisbane's leading Protestant clergymen, which called 'all our people to acknowledge their dependence upon and responsibility to the God of Battles, and to pray that His providence may make the wrath of men to praise Him'.[26] Similarly, Rev Henry Howard, minister of Adelaide's prestigious Pirie St Methodist Church, whose strident identification of the battles in France as being 'the battle of God', received an enthusiastic response wherever he spoke – although this was to prove 'disturbing to later interpreters of the church's role in society'.[27]

Equally disturbing to later generations is the fact that there seemed to be so few church leaders who were prepared to question the idea of the war being a noble crusade. There were exceptions: one was Rev Horace Crotty, rector of St Thomas' in North Sydney. In a sermon titled 'the Crisis of Nations', he spoke of the need for the churches to examine the principles behind the decision to go to war. He stressed that 'It was at such times as these that…the nation looked to its religion for light', and that 'the Church's mission was to stress first principles'.[28] The 'first

20 Kildea, 'What Price Loyalty?', 30.
21 *Tribune*, 10 October 1914, 2.
22 *Daily Telegraph*. 6 August 1914, 9.
23 McKernan, *Australian Churches at War*, 31.
24 *West Australian*, 10 August 1914, 8.
25 William Shakespeare, *Henry V*, Act 4, Scene 1; probably influenced by Exodus 15:3.
26 *Brisbane Courier*, 7 August 1914, 10.
27 Arnold D. Hunt, 'Howard, Henry (1859–1933)', *ADB* Volume 9 (Melbourne: MUP, 1983), 376.
28 *Daily Telegraph*, 3 August 1914, 7.

principles' he spoke of most certainly included the need to question whether the traditional criteria for a 'Just War' had been fulfilled.[29] The theology of the 'Just War', though a later twentieth century phenomenon, was evident in the language of most clergy, who viewed the war at least as a necessary evil, especially in the light of Germany's callous violation of Belgium's neutrality, and would have endorsed Harvey Seifert's words that 'The Christian is committed to other social goals in addition to peace. He is also concerned about justice, security and freedom'.[30]

Even so, there were people throughout the churches who, before giving their support, needed to be convinced that this war was going to be the lesser of two evils. Glen O'Brien, responding to McKernan's claim that Methodists, like the other Protestant Churches, functioned as propagandists for the British war effort, argues for a more nuanced view:

> In the early stages of the Great War, the Victoria and Tasmania Conference expressed horror and revulsion at the prospect of Christian nations at war with each other in Europe ... Though there is evidence of some support for pacifism and conscientious objection, neither of these views ever exceeded the status of a minority viewpoint. The War was seen by Methodists as a just one, fought in a righteous cause. As the conflict escalated, and it became clear that the War would last much longer than at first anticipated, the Church's early statements regarding the incompatibility of Christian faith with aggressive militarism hardened into a grim determination to win the War at all costs.[31]

---

29 The doctrine of the just war has, from the time of St Augustine (*Contra Faustum Manichaeum* 22.69–75), been the Church's attempt to set forth the minimum conditions under which Christians might reasonably participate in war. These conditions are usually defined as: 1) the war must be waged by a constituted authority; 2) the cause must be just; 3) there must be the intention of establishing good or rectifying evil; and 4) the war must be waged by proper means. See John Macquarrie, 'Just War,' in *DCE*, ed. John Macquarrie (London: SCM, 1971), 183.

30 Harvey Seifert, 'Peace and War,' in *Dictionary of Christian Ethics*, ed. John Macquarrie, (London: SCM, 1971), 248.

31 Glen O'Brien, 'The Empire's Titanic Struggle, Victorian Methodism and the Great War,' *Aldersgate Papers*, Volume 10, (September 2012), 50-70.

Response of the Other Protestant Denominations

The Other Protestant Denominations went through a similar process of adjusting to changed circumstances. Baptist leaders, such as Rev Benjamin Gawthrop, President of the Baptist Union of New South Wales, were particularly active in support of the war effort:

> What the issue will be we cannot tell, but we know this, that our conscience is clear, as that so far as Britain is concerned, our Baptist people in this Commonwealth, and throughout the Empire will be found as loyal and as ready to serve their King and country as any other section of the people.[32]

Michael Petras cites patriotism as 'unquestionably, a strong motivation', and refers to prominent Baptist ministers like Thomas Ruth of Collins Street, Melbourne, Peter Fleming of Flinders Street, Adelaide, and Stephen Sharp in New South Wales, who were all 'fervent supporters of the war effort'.[33] The *Australian Baptist* said of Fleming, who later served overseas with the YMCA as part of its ministry to the troops,[34] that he 'might well claim to have been the "recruiting sergeant" who influenced most of those brave young men to rally to the colours'.[35]

This patriotic fervour was sometimes tempered by the fear of being seen to support warmongering and the political machinations that gave rise to it. The *Australian Baptist,* just prior to the declaration of war, affirmed:

> The fact that we as Baptists have appointed chaplains to the Forces does not commit us to supporting any political view of those Forces. We may utterly hate and oppose war and yet send nurses to the front. We simply recognise that as large numbers of men go into camp their moral and religious wellbeing needs attention there.[36]

---

32 Baptist Union of NSW, *Year Book*, 1014-1915, 17.
33 Michael Petras (ed.), 'Australian Baptists and The First World War in Retrospect", in *Australian Baptists and World War I,* (Sydney: Baptist Historical Society NSW, 2009), 11.
34 The YMCA, like the Salvation Army, was active throughout the war providing welfare and recreational support to the troops.
35 *Australian Baptist,* 26 December 1916, 3.
36 *Australian Baptist,* 26 May 1914, 4.

Petras, however, argues that Australian Baptists overwhelmingly identified with the King and the Empire throughout the war. Ten days before the Gallipoli landings the New South Wales Baptist Assembly affirmed 'its unswerving loyalty to his gracious majesty the King and its undeviating loyalty to the Glorious British Empire'.[37] Five months later the Baptist Union President presented another motion of loyalty that expressed 'the Union's unswerving conviction of the absolute righteousness of the cause for which Great Britain and her allies were fighting'.[38] Even after the horrific casualty rates became apparent, 'fidelity to the King and to the Empire, remained unquestionably strong'.[39]

The Churches of Christ also showed themselves committed to Australia's support for Britain and the Empire, despite evidence of anti-militarism among some of its thinkers. Graeme Chapman notes that whereas 'in previous eras, within Churches of Christ political comment was incidental ... the Great War ... drew them out of their relative political isolation'.[40] Frederick Dunn, editor of the national journal, the *Australian Christian*, in August 1914 declared that it was 'the mission of the Church through all its agencies to deprecate the cultivation of an aggressive war spirit, and to promote the idea of arbitration as the only legitimate way in which the quarrels of nations can be settled'.[41] He continued to express confidence that the teachings of the 'Prince of Peace' would yet prevail, 'awakening the national conscience', and that 'this humanising tendency would continue to grow among civilized nations'.[42] But by mid-September, Alexander Main, the new editor of the national journal, was espousing a more realistic point of view. He dismissed Dunn's idealism, declaring that a 'one-eyed optimism cannot cheer us'. He urged his readers not 'to seek refuge in a godless pessimism' but to remember that 'that nation [Germany] ... has yet to reckon with

---

37   Baptist Union of NSW, *Executive Committee Meeting Minutes*, 15 April 1915, 137.
38   Baptist Union of NSW, *Executive Committee Meeting Minutes*, 22 September 1915, 164.
39   Petras, *Australian Baptists and World War I*, 15.
40   Graeme Chapman, *One Lord, One Faith, One Baptism: A History of Churches of Christ in Australia* (Melbourne: Vital, 1979), 113.
41   *Australian Christian* 27 August 1914, 579-580.
42   Frederick Dunn, 'Christianity and Warfare,' *Australian Christian* 27 August 1914, 579-580.

God', and that though God's desire was not being done, his will and purpose would not fail'.[43]

It is clear that most members of Churches of Christ equated God's will and purpose with loyalty to King and Empire, expressed in active support of this 'Just War'. Resolutions of loyalty were passed at both State and Federal Conferences. The first came in September 1915 when the South Australian churches expressed their 'loyalty to King and Empire in sympathy with the cause of the allies'.[44] New South Wales,[45] Queensland and Western Australian Conferences passed similar resolutions in 1916. In September 1916 the Federal Conference in Adelaide, in a strongly worded motion, also voiced its 'profound conviction in the justice of the cause of the allies'.[46] Few dissented from the view that the war was a necessary duty, and some who had previously been opponents of the war 'found their attitude changing under the pressure of imperial sentiment, which was heightened by what was written up as German barbarity'.[47]

Likewise, the Congregational Church was equally affirming of its loyalty to King and Empire. Hugh Jackson estimates that 'about one in every four adult Australian Congregational males served overseas during the Great War,'[48] and that Congregational spokesmen 'almost without exception' enthusiastically supported the Australian war effort.[49] He refers to Ethel May Gardiner who, during the second half of 1915 when a massive recruiting campaign was being conducted in Victoria, wrote to the *Victorian Independent* protesting that there was not a Congregational church where one could be sure that the 'gospel of war' would not be preached Sunday after Sunday. It was a claim the Congregational Ministers' Fraternal agreed was substantially true.[50] The New South Wales Congregational Yearbook stated that Congregational ministers who identified themselves with the war effort were no threat to

---

43 Alexander Main, 'Thy God Reigneth,' *Australian Christian* 17 September 1914, 699.
44 Chapman, *One Lord, One Faith, One Baptism*, 114.
45 *Australian Christian* 18 May 1916, 306.
46 *Australian Christian* 28 September 1916, 584.
47 Chapman, *One Lord, One Faith, One Baptism*, 114.
48 Hugh Rutherford Jackson, 'Aspects of Congregationalism in South-Eastern Australia, circa 1880–1930,' (PhD thesis, ANU, 1978), 145.
49 Jackson, 'Aspects of Congregationalism', 148.
50 *Victorian Independent,* August 1915; Jackson, 149.

church unity in that they were at one with their people. It also reported that there was 'very little patience amongst Congregationalists' with anything less than a self-righteous nationalism.[51]

There were exceptions: Rev Albert Rivett resigned his position as minister of the Whitefield Congregational Church because of growing differences with church officials over his attitude to the war.[52] Likewise Rev Thomas Bede Roseby, a life-long pacifist, sealed his fate as a dissident when he caused a riot in the Congregational Church at Orange, New South Wales. He refused to stand, take off his hat, and sing the national anthem during the church service, and was attacked by some members of the congregation including some returned soldiers. Considerable damage was done to the church building, and he was subsequently charged and fined under the War Precautions Act.[53] But he, like Rivett, was something of a lone voice in his denomination.

For the Salvation Army in Australia World War I had particular significance in shaping its unique identity and popular image. An unidentified Australian soldier in France summed up what has become an almost universal response from ex-servicemen and women:

> You cannot get away from the Salvation Army. If you are hungry, they meet you with eggs and bacon; if you are mopish they'll cheer you up with a song; if you're not doing the straight thing, they give you a rough time in the meetings; if you are put out of action by a shell, they give you a ride on a motor-car; if you 'go west' they put a marble slab to mark your resting-place.[54]

The Salvation Army's response to the war was not so much a question of patriotism and loyalty (though it would certainly have included these things) as a call to action to do what the Salvation Army always does in times of crisis. Its first published policy statement on war appeared in the *War Cry*[55] two weeks before the start of the Boer War. It stated that Salvationists must never encourage the spirit of war, never take

---

51 Congregational Union of New South Wales, *Yearbook*, 1916, 156f.
52 C.B. Schedvin, 'Rivett, Albert 1885–1934,' *ADB* Volume 11 (Melbourne: MUP, 1988), 398.
53 *Sydney Morning Herald*, 8 June 1918, 14.
54 Lindsay Cox, 'A Tale of Two Armies,' *Halleluiah!* Volume 1, Issue 3, Autumn 2008, 1.
55 *The War Cry* is the Salvation Army's monthly magazine.

sides, pray unceasingly both for peace and for Salvationists compelled to fight, and for those sent out to serve the bodies and souls of both sides.⁵⁶ David Woodbury, however, declares: 'The events of this war [World War I] were to set the [Salvation] Army on a path that would determine the image of the movement for decades to come'.⁵⁷ Australian Prime Minister, William Hughes, summed it up in a letter of thanks to the Commissioner:

> Your organisation: by its faithful, unselfish and persevering service in all climates and under all conditions, has played a great part in the victory we have achieved, and endeared itself to the hearts and minds of all Australians who went forth to fight.⁵⁸

Like the other smaller Protestant bodies, dissenting voices in the Salvation Army were relatively few and, as Reynaud observes: 'only the Quakers took a formal stand against participation in the war'.⁵⁹ Clergy and laity alike overwhelmingly shared the widespread belief of involvement in a noble cause. Pacifists were a distinct minority, and most Australian clergymen dismissed their arguments as simplistic. Nevertheless, McKernan states that the pacifists won the respect of many, 'indicating the tension inherent in the orthodox Christian response to the war'.⁶⁰ But his conclusion remains that:

> clergymen never broke themselves clear of the events, never gave themselves the opportunity to see events in perspective so that they were ever reacting rather than acting. Their initial response to the war determined their position until peace was declared ... Clergymen discussed the war in the general framework of their belief in God's oversight of the world and they decided that because God had permitted the war, he must have decided that good would come of it.⁶¹

---

56 *War Cry*, UK Edition, 30 September 1899; Cox, 'A Tale of Two Armies', 3.
57 David Woodbury, 'When the World Went to War, the Salvation Army Was There,' *Halleluiah!* Volume 1, Issue 3, Autumn 2008, 24.
58 John Bond, *The Army that went with the Boys* (Melbourne: The Salvation Army, 1919), 9.
59 Reynaud, 'Religion (Australia),' in 1914–1918, 10.
60 McKernan, *Australian Churches at War*, 144.
61 McKernan, *Australian Churches at War*, 172-173.

The Other Protestant Denominations Bid for Equal Status

It was against this background of patriotic fervour that the Other Protestant Denominations sought a greater representation in military chaplaincy. As previously noted, the chaplaincy establishment that came into effect on 2 September 1913 made allowance for District Commandants to recommend the appointment of clergy from other recognised religious bodies, who would be placed on an Unattached List.[62] The outbreak of war prompted a revision in the number of chaplains allowed for each denomination. McKernan attributes this to a decision by department officials who, assuming that the numbers of men from each church enlisting would reflect the denomination's actual size, decided that it would be fairer 'to use the 1911 census figures as a means of apportioning chaplains'.[63] Gladwin reports that this caused friction between the participating denominations. The Anglican Church in particular felt it was unfair, believing that census figures under-represented the number of Anglicans who had enlisted. His conclusion is that: 'Although it was an imperfect system that was open to unholy squabbles about money, there were few more equitable alternatives'.[64]

For the Other Protestant Denominations, though, the expansion of the Chaplains' Department to meet wartime needs opened the way for its chaplains to be posted to operational areas, and even to combat brigades. The normal practice during World War I was to appoint four chaplains to each brigade: usually two Anglicans, one Roman Catholic, and one either Presbyterian or Methodist. 'Occasionally the second Anglican was replaced by an OPD chaplain.[65] Having a team of chaplains representing the major divisions of the Christian faith in Australia posted to each brigade clearly implied that each chaplain's primary responsibility was to serve those of his own denomination. This concept was evident in the negotiations that led to the establishment of an Army Chaplains Department in 1913, and in the 1914 decision to appoint chaplains in accordance with the size of each denomination based on

---

62 Chaplains Conference Proceedings: Revision of Regulations, NAA A2023, A82/1/24.19.
63 McKernan, *Australian Churches at War,* 41.
64 Gladwin, *Captains of the Soul,* 33.
65 Reynaud, 'Religion (Australia*),*' in 1914–1918, 5-6.

the 1911 census. It demonstrates that chaplaincy was for the most part intended to be denominational chaplaincy.

Each of the four chaplains posted to a brigade, however, actually lived with[66] one or other of its battalions, and naturally tended to become associated with it rather than the brigade.[67] Although his primary role was to provide a particular denominational ministry across the whole brigade,[68] the reality of life on operational service was that each one came to be seen as chaplain to that particular battalion and all its members, Anglican, Catholic, Protestant, believer and non-believer.

Herein lies the origin of the widely accepted claim by Australian Army chaplains that it was they who first put ecumenism into practice in Australian church life. The years leading up to World War I had seen several attempts to foster closer relationships between denominations. In 1906 and 1907 the Presbyterians conducted confidential negotiations with the Anglicans on doctrine and church order, and some Protestant leaders proposed the forming of a 'United Evangelical Protestant Church which would be able to check the forces of sin and unbelief'.[69] In New South Wales and Western Australia the Baptists and Churches of Christ set up committees to explore the possibility of union between them.[70] Neither endeavour succeeded, and an attempt to promote union between Congregationalists, Methodists and Presbyterians was finally abandoned in 1920.[71]

Douglas Abbott, however, refutes the claim by the Chaplains General that they were at the forefront of ecumenical change, describing it as 'shadow rather than substance'. The substance of it came from the chaplains in the trenches. Though not concerned with matters of doctrine and church order, it went beyond such tokens as 'agreeing to share in silent prayers with clergy of other denominations.'[72] It was the practical reality of ministering to men of all faiths – or no faith – amidst the

---

66 The military term is 'detached to'.
67 An Australian brigade during World War I comprised four infantry battalions plus support units.
68 McKernan, *Australian Churches at War*, 41.
69 Ian Breward, *A History of the Australian Churches*, (Sydney: Allen & Unwin, 1993), 99.
70 Chapman, *One Lord, One Faith, One Baptism*, 107.
71 Ian Breward, *A History of the Australian Churches*, 100.
72 Abbott, *In This Sign Conquer*, 10.

horrors of industrialised slaughter. Regardless of the denominational structure that had been imposed on the military, in the trenches of Gallipoli and the Western Front the seeds of practical ecumenism were sown.

Over the four years of the war four hundred and fourteen Australian clergymen served as chaplains with the Australian Imperial Force: one hundred and seventy-five Anglicans, eighty-six Catholics, seventy Presbyterians, fifty-four Methodists, and twenty-seven from the Other Protestant Denominations. The system of denominationally based chaplaincy meant that many of the latter were posted to base areas 'where they would meet a wider number of their adherents' rather than to operational brigades, but this was by no means universal.[73]

However, the story of the struggle by the Other Protestant Denominations to be seen as legitimate providers of effective chaplains goes beyond mere statistics. It has overtones of denominational pride, protectionism, and a failure to recognise the need for a more ecumenical form of chaplaincy in an environment where most of the soldiers really had no commitment to any denomination. The typical Australian soldier was not religious in the sense of being an active churchgoer. Gammage, somewhat cynically, refers to him as avoiding church parades, 'or if he could not avoid them, he tended to show sudden enthusiasm for whichever denomination worshipped within easiest marching distance'.[74] Nevertheless, as Stuart Piggin and Robert Linder assert, a more nuanced view of Australian soldiers' religiosity is emerging, revealing much more interest in religious issues and appreciation of chaplains' ministry than people like Gammage assumed.[75] But it tended not to be primarily focused denominationally.

In a bid to gain equal status with the major churches the Council of Churches in Victoria, representing the Other Protestant Denominations, in July 1915 petitioned the Minister of Defence for the appointment of their own Chaplain General.[76] The request was passed down to the

---

73 McKernan, *Australian Churches at War*, 41; Patrick Porter, 'The Sacred Service: Australian Military Chaplains in the Great War,' *War and Society*, Volume 20, (2002), 23.
74 Gammage, *The Broken Years*, xiv.
75 Stuart Piggin and Robert Linder, *Attending to the National Soul* (Melbourne: Monash University Publishing, 2020), 61-62.
76 Chaplain General – Other Protestant Denominations, NAA A2023, A82/1/191; Gladwin, *Captains of the Soul*, 33.

Army's Adjutant General who consulted the existing Chaplains General. They were unanimous in their opposition. Presbyterian Chaplain General John Laurence Rentoul summed up their attitude declaring that the petition was 'unreasonable, unnecessary and undesirable'. He further described the Other Protestant Denominations as 'an aggregate of quite non-coherent and non-corporate factors having no solidarity or unified administration and responsibility'. His conclusion was that an OPD Chaplain General 'would be responsible to nobody' and that such an appointment would 'serve to make the office and the title of Chaplain General belittled, unmeaning and worthless'.[77]

The Adjutant General accepted their advice and decided not to proceed with the appointment. He gave his reasons in a letter to the Commandant 4th Military District on 12 January 1916:

> So important a position as Chaplain General should represent a reasonably large proportion of the community; and that there is no more affinity between Baptists, Congregationalists and Churches of Christ than between Anglicans, Presbyterians and Methodists. The appointment of a Senior Chaplain in each Military District to represent the Baptist, Congregational and Churches of Christ affords ample distinction to those bodies, and under the present system no injustice is done to them.[78]

The last sentence makes it clear that, in addition to the Senior Chaplains of the major churches, by then each of the six Military Districts had an OPD Senior Chaplain. It is also noteworthy that the Salvation Army was not included in the Adjutant General's mention of the members of the group making the petition. This may be because the Salvation Army had already sought special representation at Senior Chaplain level. On 14 August 1914 the Chief Secretary of the Salvation Army wrote to the Department of the Army suggesting that there ought to be one Salvation Army Senior Chaplain appointed for the whole of the

---

77 Chaplain General – Other Protestant Denominations, NAA A2023, A82/1/191; Abbott, *In This Sign Conquer*, 54.
78 Chaplain General – Other Protestant Denominations, NAA A2023, A82/1/191; Abbott, *In This Sign Conquer*, 55.

Commonwealth.⁷⁹ It appears that the Salvation Army may have hoped to gain its own recognition and status in the overall establishment.

This raised the question of whether Salvation Army officers were actually qualified to become chaplains. William Booth – founder of the Salvation Army – never intended that the movement he established should be a church but rather a mission to the poor and working classes.⁸⁰ It did not administer the sacraments and chose not to ordain clergy; instead, using military terminology, it 'commissioned' its officers. Furthermore, its officers did not have an equivalent level of theological education to other Protestant ministers.⁸¹

Abbott reports: 'The request [for its own Senior Chaplain] met a mixed reception from existing Senior Chaplains in Military Districts'.⁸² Rev Albert Thomas Holden, Methodist Chaplain General, wrote to the Adjutant General on 8 January 1914 pointing out that the Salvation Army only represented 0.7% of the Australian population and that 'such a small proportion of the men ... might well be regarded as having their spiritual needs attended to by Protestant Chaplains'.⁸³ Consequently, the request for a Salvation Army Senior Chaplain was declined, and though Salvation Army Chaplains were soon appointed, they worked under the oversight of the OPD Senior Chaplain.

The Status Quo Maintained

It was inevitable that the Other Protestant Denominations' request for equal standing with the major churches was denied. Sectarianism was a major factor in the religious life of Australia at that time. Ecumenism, which David Wright says, 'loomed large in 20th Century Christianity ... after the stimulus of the Edinburgh Missionary Conference (1910)',⁸⁴

---

79 Senior Chaplain Salvation Army, NAA A2023, A82/1/174; Abbott, *In This Sign Conquer*, 56.
80 Its original name was the *East London Revival Society*, and afterwards the *East London Christian Mission*.
81 Gladwin, *Captains of the Soul*, 34.
82 Abbott, *In This Sign Conquer*, 56.
83 Appointment of Chaplains to Salvation Army and other Denominations, NAA A2023, A82/1/157.
84 David F. Wright, 'Ecumenical Movement,' in *New Dictionary of Theology*, eds. Sinclair B. Ferguson and David F. Wright (Leicester: IVP, 1988), 219.

and the idea of generic or trans-denominational chaplaincy, were far from the minds of those religious leaders who negotiated the establishment of Australia's military chaplaincy organisations.

It even appears to have been absent in the thinking of some representatives of the Other Protestant Denominations, despite their concerns about being marginalised on the basis of statistics. Frederick James Miles, whom Petras describes as 'one of the most outstanding chaplains of the First World War,'[85] was one of the first chaplains appointed when war broke out. Miles already had a military background, having served in the British Army. As a Baptist minister he had been appointed as a chaplain in the Citizens Military Force and eventually became one of the Senior Chaplains of the 3rd Military District.[86]

He sailed from Melbourne with the 6th Infantry Battalion on 19 October 1914, and in one of his first letters, having tabulated the attestation papers of the religions of the troops on board, he commented on the absurdity of having a Salvation Army chaplain for this force. 'Most of the 2,000 men on this boat,' he said, 'are drawn from the Melbourne districts where the [Salvation] army is strongest, yet there are only two soldiers of the Salvation Army on the ship'.[87]

This, of course, raises the question of how many Baptists were aboard the ship. According to the 1911 census there were 26,665 members of the Salvation Army in Australia, and 97,074 Baptists.[88] So, statistically, it is possible that there may only have been seven or eight Baptists among the 2000 soldiers.[89] If one uses declared religious affiliation as the basis for appointing chaplains Miles should probably not have been there either. Fortunately, the rationale behind such strict adherence to denominational proportions did not prevent either of these two chaplains from serving. Miles went on to become the OPD Senior Chaplain on the Western Front, while the Salvation Army chaplain was none

---

85 Petras, *Australian Baptists and World War I*, 23.
86 His name appears on a letter written by the Assistant Adjutant General, dated 11 July 1911, along with that of a Congregational Senior Chaplain, Rev E. Davies; Chaplains – Conference Proceedings, NAA A2023, A82/1/24.
87 Petras, *Australian Baptists and World War I*, 23.
88 Commonwealth of Australia, Census, 3 April 1911.
89 Miles did not mention the number of Baptists aboard.

other than the redoubtable William 'Fighting Mac' McKenzie,[90] whom McKernan describes as 'one of the most popular men in the AIF'. In later years his hand was often seen to be bleeding after an Anzac Day march, 'so many men being anxious to grasp and greet him'.[91]

Nonetheless, the concept of people of all denominations working together in mission on the basis of spiritual and pastoral need, rather than denominational protectionism, was not unknown. Interdenominational mission work was by then well established in Australian evangelical churches, evidenced by the numerous missionary societies that drew their workers and their financial support from such congregations. Stephen Neill traces its origin to the foundation of the China Inland Mission[92] by James Hudson Taylor, who in 1865 'was led to undertake single-handed the foundation of what for a time was the largest mission in the world'. Neill remarks that its influence was such that 'During the last third of the century, missionary societies of every conceivable kind, and other organizations such as the YMCA extended themselves throughout the length and breadth of the land'.[93]

The interdenominational ethos, however, was unable to penetrate the centuries old traditions of ordained exclusiveness, especially those claiming apostolic succession. Consequently, there really was little possibility that the leaders of the major churches would be prepared to accept the Other Protestant Denominations as anything more than support players in the developing drama. This is evident in an article published in the *Australian Baptist,* complaining about a report that appeared in the West Australian daily press. It concerned comments made by Anglican Archbishop Charles Owen Riley that 'he could not sanction on behalf of the Church of England that a Baptist or Church of Christ minister should be officially in charge of Church of England men'. The archbishop, while listing the numbers of Anglican, Roman Catholic, Presbyterian and Methodist chaplains, said nothing about OPD chaplains, thereby 'disposing of Baptists, Congregationalists, and Churches of Christ chaplains.' The writer concluded by saying:

---

90   Petras, *Australian Baptists and World War I,* 23.
91   McKernan, *Padre* (Sydney: Allen & Unwin, 1986), 3.
92   Now known as the Overseas Missionary Fellowship.
93   Stephen Neill, *A History of Christian Missions* (Melbourne: Penguin, 1964), 335-336.

> At a time like this, when a common peril and a common need should be drawing every citizen of the Empire into closer bonds of brotherhood, no Baptist voice, it is certain, will be raised against an Anglican chaplain offering spiritual ministration to Baptist soldiers. Why cannot Archbishop Riley rise superior to his ecclesiastical prejudices and adopt the same attitude of happy Christian tolerance towards Baptists?[94]

In fairness the Chaplains General, while resistant to a growing involvement by the Other Protestant Denominations, might have been motivated by concerns for maintaining denominational prestige. Deeply held theological and ecclesiastical beliefs would have also been part of the mix. The importance of sacramental ministry and church tradition as channels of grace and encounter with God would have weighed heavily on the hearts and minds of those from a High Anglican and Roman Catholic ecclesiology. The Protestant Reformation and four centuries of European Church history, with all the horrors of religious wars and bitterness of sectarian persecution, had exposed profound divisions in Christendom, both in ecclesiology and theology. All these lay behind this tentative coming together of Christian traditions, and it is understandable that, in the context of the day, leaders of the major churches – especially those of episcopal traditions like Archbishop Riley – might be less enthusiastic than their evangelical 'free-church' brethren in favour of trans-denominational pragmatism.

Nevertheless, there was some truth in the Presbyterian Chaplain General's previously mentioned dismissal of the Other Protestant Denominations as 'an aggregate of quite non-coherent and non-corporate factors having no solidarity or unified organic administration and responsibility'. Each of the member churches had its own structures of leadership and administration, systems of ordaining or accrediting ministers, methods of administering pastoral oversight and discipline, and traditions that shaped its theological and ecclesiological ethos. The Baptists and Churches of Christ were strongly committed to believer's baptism, while the Congregationalists were paedobaptists, and the Salvation Army didn't baptise at all. The Churches of Christ were wary of

---

94 *Australian Baptist*, 17 August 1915. 8

the liberalism of the Congregational church, especially with regard to 'social and humanitarian subjects and questions of politics and higher criticism,'[95] and it is likely that the Baptists, who shared many similarities with Churches of Christ, and the fervently evangelistic Salvation Army, felt the same way.

As for the Salvation Army, it seems, like the major churches, the other three member churches were not sure how to categorise it. Was it a church or a mission? There was universal admiration for its welfare work which many saw as its real ministry. Although admiring the Salvation Army for its benevolent work, the general opinion within Churches of Christ was that without such work, having abandoned the sacraments, 'it would have no reason for existence as a separate body'.[96] Furthermore, while all four denominations were part of worldwide organisations, only the Salvation Army was subject to an international hierarchy.[97] In these respects it is easy to understand the Presbyterian Chaplain General's negative perception of the group.

Even so, the Other Protestant Denominations succeeded in fulfilling the Army's requirements to provide the required number of chaplains in proportion to their share of the overall population.[98] It accomplished this through State-based boards,[99] which nominated potential chaplains to local Military District Commandants.[100] Prior to this several ministers from its member denominations had been appointed to CMF units before 1913. Mention has already been made of Metters (Baptist) having begun his chaplaincy service in 1906.[101] Miles (Baptist) and Davies (Congregational) had been serving as senior chaplains for 3rd Military District as early as 1911. Walden (Churches of Christ) was appointed chaplain to the 4th Military District in September 1913.[102] Prior to the

---

95 Chapman, *One Lord, One Faith, One Baptism*, 106.
96 Chapman, *One Lord, One Faith, One Baptism*, 107.
97 Breward, *A History of the Australian Churches*, 219.
98 From a total population of 4,455,005 people, 236,533 or 5% were adherents of the Other Protestant Denominations, and their 27 chaplains comprised approximately 6.5% of the 414 chaplains who served in the AIF; McKernan, *Australian Churches at War*, 41; Commonwealth of Australia Census, 3 April 1911.
99 The four member denominations were already associated through State councils of churches.
100 Congregational Union of NSW Yearbook, 1916, 65.
101 *Australian Baptist*, 31 July 1917, 3.
102 Nutt, 'Military Chaplains,' 20. Walden was the first chaplain from Churches of Christ.

outbreak of war they were joined by Lamb (Baptist) to the 2nd Military District in May 1914, plus six other Baptists who ministered to youths attending periods of compulsory training.[103]

Despite their criticism of the seemingly casual nature of the Other Protestant Denominations association, it is significant to note that the four existing Chaplains General do not appear to have had a close working relationship either. There is no evidence that they ever met together in conference, (Archbishop Riley was located in Perth, which was then a three-day train journey from Melbourne), and each of them corresponded individually with the Adjutant General, and through him to the Secretary of the Department of Defence.[104]

In summary, until 1914 the Other Protestant Denominations had a small involvement in Australian Army chaplaincy and none at all in the Royal Australian Navy, which continued to be the domain of the major churches until 1969. It was the declaration of war and the outpouring of patriotic fervour that changed all that. Baptists, Congregationalists, Churches of Christ, and the Salvation Army all pledged their loyalty and support to King and country, their young men joined the thousands that flocked to the recruiting centres, and the governing bodies of those denominations offered their ministers to serve as chaplains. Their request for equal status in the military chaplaincy organisation was largely foiled by the influence of the existing Chaplains General, who eventually agreed that chaplaincy numbers should be based on the relative size of denominational memberships, as reported in the 1911 national census.

It was this, as well as the 1913 agreement, that opened the door for the Other Protestant Denominations to nominate the twenty-seven clergy who served overseas with the 1st AIF and as 'voyage only' chaplains on troop ships, along with others who ministered in military camps and hospitals within Australia. The 1914 decision to appoint chaplains on the basis of census figures now guaranteed their right to proportional representation. The next four years proved their worth as they joined chaplains of the major churches in helping keep faith, hope, and love alive amidst the horrors of industrialised war.

---

103 *Australian Baptist*, 29 December 1914, 4.
104 Abbott, *In This Sign Conquer*, 57.

CHAPTER THREE

World War I: The Weathering of the Stone

SIX DAYS BEFORE AUSTRALIA declared war Andrew Fisher, leader of the Federal Opposition, declared: 'Australians will stand beside the mother country to help and defend her to our last man and our last shilling'.[1] It was a sentiment the Government shared because four days later it offered to place the Australian Navy under the British Admiralty, and to despatch 'a force of 20,000 men ... to any destination desired by the Home Government'.[2] The British gladly accepted the offer, and the *Sydney Morning Herald* next day reported that the Governor-General had received messages from both the British Government and the King expressing their appreciation.[3]

Brigadier General William Bridges was promoted to Major General and ordered to raise this force, which he designated the Australian Imperial Force.[4] Men flooded the recruiting depots, and many clergymen, equally enthusiastic, approached their bishops or governing

---

1   Part of an election speech by Andrew Fisher at Colac, Victoria, 31 July 1914; *The Age*. 1 August 1914, 15.
2   Charles Bean, *Anzac to Amiens* (Canberra: Australian War Memorial, 1983), 25.
3   *Sydney Morning Herald,* 6 August 1914, 1.
4   Chris Coulthard-Clark, 'Major-General Sir William Bridges: Australia's First Field Commander,' in *The Commanders, Australian Military Leadership in the Twentieth Century,* ed. D.M. Horner (Sydney: Allen & Unwin, 1984), 18.

bodies, hoping to be appointed as chaplains. However, the number of clergy applicants far exceeded the positions available, and many were rejected.

Michael McKernan describes the first chaplains selected as 'a curious bunch'. Some were Boer War veterans while others had no military experience, and many of them were quite old.[5] Apart from Roman Catholics, most of whom were over forty – the first one nominated by the Archbishop of Melbourne was over sixty[6] – their average age was between thirty and forty. Thus, they were younger than most civilian clergymen[7] but older than most soldiers, the majority of whom were under twenty-nine.[8] Tom Johnstone concludes that many of the Catholics were 'over age and ill-suited for the rigours of campaigning' and marvels that 'so many of them lasted in the field as long as they did'.[9]

Among those over the age of forty who did last the distance were eight OPD chaplains, including Frederick Miles (Baptist), who was forty-five when he enlisted; William McKenzie (Salvation Army), who was forty-four; and George Walden (Churches of Christ), who was fifty-three.[10] Between them they served for fourteen years and three months in the 1st AIF.[11]

The problem of too many volunteers was partially remedied by a proposal that 'honorary chaplains should be allowed to serve on transports, without pay or rank'.[12] This enabled the Chaplains General to appoint both 'continuous service' chaplains and 'voyage only' chaplains who would serve on the troopships, then return as soon as practicable as hospital ship chaplains.[13]

---

5   Michael McKernan, *Padre* (Sydney: Allen & Unwin, 1986), 1.
6   Tom Johnstone, *The Cross of Anzac, Australian Catholic Service Chaplains* (Brisbane: Church Archivists' Press, 2003),19.
7   Michael Gladwin, *Captains of the Soul. A History of Australian Army Chaplains* (Sydney: Big Sky, 2013), 35.
8   Jeffrey Grey, *A Military History of Australia* (Melbourne: CUP, 2008), 89.
9   Johnstone, *The Cross of Anzac*, 34.
10 Chaplains [alphabetical], AWM 8 6/6/3, 1-16.
11 Miles F.G., McKenzie W., Walden G.T, Military Records, NAA, B2455
12 Michael McKernan, *Australian Churches at War; Attitudes and Activities of the Major Churches 1914–1918* (Sydney: Catholic Theological Faculty and Australian War Memorial, 1980), 40.
13 Douglas Abbott, 'In This Sign Conquer: The Chaplains General of the Australian Army, 1913–1981'. Unpublished manuscript, 1995, 58.

Many were still disappointed, and instead enlisted as ordinary soldiers. The *Australian Baptist* reported that despite the exemption from military service granted to ministers and theological students,[14] by 1916 'many, if not all, had already joined up'.[15] Neither was this peculiar to Baptists. Five students from the Churches of Christ College of the Bible in Melbourne enlisted, two of whom were killed. One who survived, Daniel Wakeley, was awarded the Military Medal.[16] Several Congregationalist students and clergy also enlisted, two of whom were killed.[17]

Thomas Perkins (Congregational) who, though already serving as a chaplain in the CMF, chose specifically to enlist as an ordinary soldier. The *Bendigoan* said of him:

> He has closely identified himself with the life of the soldiers ... he has risen with the men, been with them on parade, participated with them in their physical exercises and marches, and even helped in trench digging. His decision to become a private occasioned surprise ... but he will have great opportunities for doing the work of a Christian minister among the men ... when on active service.[18]

Similarly, Arthur Forbes (Churches of Christ), who was awarded the Distinguished Conduct Medal during the Boer War,[19] secured an appointment as a CMF chaplain in March 1915. Doubting that he'd get an opportunity to serve with the AIF, he enlisted in a medium trench-mortar battery, rising to the rank of sergeant. However, prior to embarkation he was offered an appointment as an AIF chaplain.[20]

---

14 Commonwealth of Australia *Defence Act, 1903,* Section 61A, 110.
15 *Australian Baptist,* 31 October 1916, 6; Michael Petras, 'Australian Baptists and The First World War in Retrospect, in *Australian Baptists and World War I,* ed. Michael Petras (Sydney: Baptist Historical Society of NSW, 2009) 17.
16 *Australian Christian,* 18 May 1916, 277.
17 John Garrett, and L.W. Farr, *Camden College: A Centenary History* (Sydney: Glebe, 1964), 36-37.
18 *Bendigoan,* 30 March 1916, 25.
19 The second highest award for bravery in the Field after the Victoria Cross.
20 Dennis Nutt. 'Military Chaplains: For Service of our Soldiers', *AACJ* 27, 2016, 27; Arthur Bottrell, 'Forbes, Arthur Edward (1881–1946)', *ADB* Volume 8 (Melbourne: MUP, 1981), 539.

The Other Protestant Denominations Contingent

Of the twenty-seven OPD chaplains who went on active service, six were classified 'voyage only' and twenty-one 'continuous service'. Eleven were Congregationalists, including two 'voyage only' chaplains; nine were Baptists, four being 'voyage only'; five were from Churches of Christ and two from the Salvation Army.[21] Two 'voyage only' chaplains, Alfred Austin (Congregational) and George MacKay (Baptist), held the rank of Chaplain Class 1. All the others initially held the rank of Chaplain Class 4.[22]

The only one with active service chaplaincy experience was Donald McNicol (Baptist), who served as chaplain with the Seaforth Highlanders during the Boer War and reported having come under fire on twenty-two separate occasions.[23] George Cuttriss (Churches of Christ), like Forbes, also served in South Africa, but as a private in the New Zealand Mounted Rifles.[24] The remainder, apart from Miles who had been a soldier in the British Army, went to war with little more than a rudimentary knowledge of military life, gained by attendance at annual camps as CMF chaplains, or as cadets.[25]

Despite the Presbyterian Chaplain General's dismissive reference to them as an 'aggregate of quite non-coherent and non-corporate factors …,' they proved to be a very effective component of AIF chaplaincy in every major theatre of the war. As it is not feasible to consider all twenty-seven chaplains, four of the most outstanding will serve as illustrative of the rest – Frederick Miles, George Walden, William McKenzie and Ashley Teece (Congregational) – one from each of the Other Protestant Denominations religious traditions. In addition to their representative nature, these were chosen because of the high regard in which they were held within their churches and the AIF; their diaries and prolific letters; the length of time they served on operations,

---

21 Chaplains [alphabetical], AWM 8 6/6/3, 1-16.
22 The Chaplains Department followed the practice of its British parent in classifying chaplains into four classes: Class 1 corresponding to the rank of Colonel; Class 2, Lieutenant Colonel; Class 3, Major: and Class 4, Captain.
23 *Weekly Times,* 26 June 1915, 8.
24 Nutt, 'Military Chaplains', 22.
25 Gladwin, *Captains of the Soul,* 104.

which was far in excess of what was required for chaplains, two of them becoming the longest serving of all AIF chaplains; and because they were among the most highly decorated of all AIF chaplains. Moreover, Miles, McKenzie and Walden served both at Gallipoli and on the Western Front.

Miles, who prior to the declaration of war had been a Senior Chaplain for 3rd Military District, actually had his rank reduced when transferring to the AIF, while Austin and MacKay were promoted. Miles said of this:

> When World War One broke out it had been decided that the Senior Chaplain of each Denomination in each state should be made a Chaplain First Class with the rank of Colonel. Some of my brethren had already received their promotions, but in Victoria, my promotion had not come through ... I went on active service as a Captain while other men who had never done a day's work in the forces went out as Colonels. We were paid according to rank so I suffered a loss of several hundred pounds before finances were equalised and all Padre's [sic] received the same pay in the AIF.[26]

Miles, McKenzie, and Walden served in the Gallipoli Campaign along with Theodore Robertson (Congregational), who soon after returned to Australia medically unfit, where he continued to serve.[27] Miles, McKenzie and Walden then went on to the Western Front, as did all the other 'continuous service' chaplains except for three Congregationalists who remained in Egypt and the Sinai. One of them, John Dempsey, died of septic pneumonia on 13 June 1917.[28] He was the only OPD chaplain to die on active service.

Miles, McKenzie, Walden and Teece all received awards for gallantry and distinguished service.[29] Miles was awarded the DSO, OBE, and was

---

26 Miles F.G. Military Record, NAA, B2455, 103; David Harrower, 'Senior Chaplain; Frederick James Miles, DSO, OBE, VD, MID' in Harrower Collection, 9th Infantry Brigade, AIF Online Research Library, 17April 2017.
27 Robertson T.G Military Record, NAA, B2455, 35.
28 Dempsey J. Military Record, NAA, B2455, 14.
29 Miles F.G. Military Record, NAA, B2455, 9-12; McKenzie W. Military Record, B2455, 9-12; Walden G.T. Military Record, NAA, B2455 23; Teece Ashley, B2455, 16, 33.

Mentioned in Despatches. McKenzie and Teece both received an MC and with Walden were also Mentioned in Despatches.[30] In addition to these awards, Chaplain Henry Procter (Churches of Christ) was noted for an act of bravery as his troopship was leaving the wharf at Port Melbourne:

> Mr H.A. Procter was the gallant chaplain, who recently assisted to rescue a young woman at Port Melbourne, when she was accidentally pushed over the pier by the crowd, as a troopship was leaving. Mr Procter was on the vessel as a chaplain for the Churches of Christ and is going to the Front. He dived from the second deck, and helped the young woman's lover and another soldier to save her.[31]

Miles and McKenzie were among the very first AIF chaplains appointed. Miles enlisted on 14 September 1914 and embarked from Melbourne with the 6th Battalion[32] on 19 October aboard the *Hororata*.[33] McKenzie enlisted on 25 September and embarked from Sydney with the 4th Battalion on 18 October.[34] The Chaplain Embarkation List places him aboard the *Suffolk*, but McKenzie in his diary records that he sailed on the *Euripides*.[35]

McKenzie's diary gives an insight into the devotion to duty of those men, and the personal cost to their families. It records how, having just arrived in Melbourne from nine days in Northeast Victoria, he was summoned to meet the Salvation Army Commissioner about his desire to become a chaplain. He caught the train back to his home in Bendigo, where he had tea at 9 pm then at 9.15 pm told his wife what he was about to do. She responded, 'in a very fine soldierly way' and said, 'if you feel it is the right thing to do I have no objection to raise and may God go with you and take care of you'. They then worked all night to get him packed, and he left next morning to complete enlistment formalities in Melbourne. At 5 pm he caught the train to Sydney, where

---

30 Distinguished Service Order, Officer of the Order of the British Empire, and Military Cross.
31 *Weekly Times*, 26 May 1917, 8.
32 Miles FG. Military Record, NAA, B2455, 32.
33 Chaplains [alphabetical], AWM 8, 6/6/3, 9.
34 McKenzie W. Military Record, NAA B2455, 4.
35 Chaplains [Alphabetical], AWM 8 6/6/3; William McKenzie, *Diary*, AWM 2019.22.2, 4-6.

he joined his unit and began his chaplaincy work, departing with them two weeks later.[36]

Charles Bean, the AIF's official historian, describes the departure of the first Anzac convoy thus: 'With the British cruiser *Minotaur* five miles ahead, the *Ibuki*[37] and *Melbourne* four miles out on either beam, and the *Sydney* far astern, the 38 transports headed for Suez *en route* to England'.[38] Andrew 'Banjo' Paterson who was with the convoy wrote: 'Thirty thousand fighting men, representing Australasia, are under way for the great war ... it is the most wonderful sight that an Australian ever saw'.[39] Fifteen Australian chaplains sailed with them, including Miles and McKenzie,[40] and pioneered an ethos that came to characterise OPD chaplains. It can be summarised in five characteristics.

Incarnational Ministry

Incarnational Ministry can be defined as immersing oneself into a local culture and becoming Jesus to that culture. It seeks to replace ministry that is remote from people with one that is close and personal. The love of God and the gospel of Christ are 'incarnated' or embodied by the person ministering. Just as the Son of God took on human flesh and came into our world, ministers live into the culture to which they are ministering and 'become Jesus' within it.

This concept is inextricably bound to the Christian doctrine of the Incarnation of Christ, which may be defined as 'the act of God the Son whereby he took to himself a human nature',[41] and lived among humankind, bringing comfort, healing and salvation, while himself subject to the restrictions, joys and sorrows of life in this world. The emphasis in incarnational ministry is on being engaged with people and living a life of Christlikeness. It is a theme that has long dominated pastoral

---

36 McKenzie, *Diary*, 1.
37 A Japanese cruiser.
38 Bean, *Anzac to Amiens*, 48.
39 *Sydney Morning Herald*, 8 December 1914, 8.
40 Chaplains [alphabetical], AWM 8 6/6/3, 1-16.
41 Wayne Grudem, *Systematic Theology, An Introduction to Biblical Doctrine* (London: Inter-Varsity Press, 2016), 543.

theology, expounded by such notables as Thomas à Kempis in *The Imitation of Christ*, which 'portrayed Christ as an example to us of suffering, self-deprivation, humility, patience, solitude and brokenness of will'. It has been re-emphasized in Liberation Theology and its assertion that following Jesus 'means putting Utopia into practice – [which] will involve conflict, sacrificing love, hope, faith, cross and resurrection.'[42]

More than anything else this was the key that opened the door to the soldiers' affection and respect. Despite Bill Gammage's conclusion that the average Australian soldier 'distrusted chaplains and sometimes detested them' – a claim that Daniel Reynaud dismisses as based on incomplete evidence[43] - he noted that there were exceptions 'who showed themselves brave under fire ... taught by practice and example and were among the most respected in the AIF'.[44]

Miles, McKenzie, Walden and Teece were foremost among these.[45] McKernan observed that the 'popular chaplains were those who shared the risks with the troops'. He cites McKenzie as one of those who 'won admiration because they went everywhere with their men'.[46] The same was manifestly true of Miles, Walden, and Teece. The tribute paid to McKenzie by one unnamed soldier might well be applied to many others also: 'Captain McKenzie is not a "dug-out" chaplain ... He is where the shells fall thickest, and the cries for help are most numerous. That is where we always find him'. [47]

Their concept of incarnational ministry developed in the crucible of active service. Like the soldiers, they knew that the real test would come on the battlefield. It began – ironically – on April Fool's Day. Bean records: 'Then on April 1st, like a bolt from the blue, came the order:

---

42 P.J.H. Adam, 'Jesus,' in *New Dictionary of Christian Ethics and Pastoral Theology*, ed. David J. Atkinson and David H. Field (Leicester: IVP, 1995), 508.

43 Daniel Reynaud, *Anzac Spirituality* (Melbourne: Australian Scholarly Publishing, 2018), 11-13.

44 Bill Gammage, *The Broken Years, Australian Soldiers in the Great War* (Melbourne: Penguin, 1982), xiv-xv.

45 Reynaud questions Gammage's view, as does Gladwin, and 'found that chaplains, far from Gammage's pessimistic view of them being distrusted by the soldiers, were frequently held in high regard; Reynaud, 'Religion (Australia),' 7.

46 McKernan, *Australian Churches at War*, 134-135. McKernan, M. 'McKenzie, William (1869–1947),' *ADB*, Volume 10 (Melbourne: MUP, 1986), 305-306.

47 David Woodbury, 'Do You Think I'm Afraid to Die with You', *Halleluiah*, Volume 1, Issue 3, Autumn, 2008, 14.

"All leave stopped!" The 1st Australian Division was ordered to the front.[48] Their first destination was the island of Lemnos. Miles describes his initiation as a front-line chaplain thus:

> On the ship coming from Egypt, and while at the rendezvous at Lemnos, the services were most impressive, and deepened intensity characterised them. Several conversions took place. I was privileged to serve other troops also while at Lemnos. On the first day of landing chaplains were forbidden to land, and I made a trip to Alexandria with wounded ... We were the closest inshore at the landing (600 yards), were fired on five times, received wounded back in the boats ... had the first death and the first Turkish prisoner.[49]

Miles worked for three days and two nights in an improvised operating theatre, then returned to Gallipoli where he landed and re-joined his brigade. In a graphic account of that landing he records:

> We had to climb down the ship's rope ladder into our boat ... a tug took us within 100 yards of the beach and we had to row the rest of the way. The shrapnel was bursting all round us, also machine guns & rifle shot. We lost a lot of men before we landed, but our boat got ashore safely ... As the boys took off inland, I immediately started to attend the many wounded & dying that were now gathering on the beach.[50]

His daily routine then was to move up to the firing line and support trenches, 'visiting the men and giving them a word of cheer, at times helping the wounded'. He held Church parades behind the line on Sunday mornings, with Holy Communion following, and voluntary services at night, 'weather and shells permitting', as well as camp singsongs and lectures. He took his turn with his fellow chaplains to visit the hospitals at Lemnos and made two trips with hospital ships carrying thousands of wounded to England. Through all this he felt his relationship with the soldiers deepening:

---

48 Bean, *Anzac to Amiens*, 75.
49 *Australian Baptist*, 25 January 1916, 11.
50 David Harrower, 'Senior Chaplain; Frederick James Miles, DSO, OBE, VD, MID'.

> I got closer to the men than at any previous time. Many decisions[51] were registered ... Whatever has had to be done, God has given me grace and physical and mental strength to do.[52]

Miles was probably the first Australian chaplain to be wounded. He was hit in the right eyebrow by a piece of shrapnel when a shell burst in his dugout.[53] The following day he led a church parade where the men sat close to the dug-outs ready to dive for safety. A 'glee party'[54] led the singing and afterwards more than fifty men 'filed up a narrow ravine' to a safer spot to observe the Lord's Supper. One can only imagine what it must have been like to attempt to lead men in lifting their hearts in praise to God while knowing that at any moment death or mutilation might rain down upon them. That evening, after the shelling had stopped, he held a concert with solos and hymns chosen by the men and led by the 'glee party'. The following morning, he baptised two soldiers in the Aegean Sea. Their Commanding Officer was sufficiently impressed that he took photographs of the event.[55]

McKenzie also sailed with his battalion to Lemnos, arriving on 7 April, where he managed to get on shore 'to purchase food for the boys on the ship'. While in the harbour he held several concerts, officiated at church parades and led voluntary worship services. Then, at midnight on 24 April, their darkened ship set sail for Gallipoli.[56] His diary entry for 25 April declared it 'a most memorable day'. The 3rd Brigade landed at 4 am and the Turkish guns opened fire at 4.50 am. A line of seven warships, including *Queen Elizabeth*, then bombarded Turkish positions. He described it as 'a terrible day for the men ... continuing without intermission from dawn to nightfall'. However, it was clearly not a 'terrible day' for him. His next words give a fascinating, if unexpected, insight into his character:

---

51 Evangelical terminology for expressions of personal commitment to Christ.
52 *Australian Baptist*, 25 January 1916, 11.
53 Miles Frederick, NAA B2455, 28; Petras, *Australian Baptists and World War 1*, 23.
54 'A group for singing choral music'. Macquarie Dictionary (Sydney: Macquarie Library, 1981), 756.
55 *Australian Baptist*, 24 August 1915, 3.
56 McKenzie, *Diary*, 58.

> It is remarkable to relate that when I saw the shells dropping all around us and the rifles speaking and guns going I just felt in a most gleeful happy mood. All fear vanished and I revelled in it and longed to get at them. I can never forget the sight and I am glad to have been in it.[57]

Higher command had determined that every available space on the boats was needed for fighting soldiers, and therefore chaplains were prevented from landing with the first assault – though Chaplain Fahey (Catholic) managed to evade the order, believing it was his duty to be with the men.[58] In the days that followed several other chaplains went on shore and the rest followed in mid-May.[59] McKenzie disembarked on 10 May. His diary entry for that day records:

> I had a very narrow squeak (escape), got covered with earth when I threw myself on the road for safety. A number of shells fell near us. However, I got safely up the hill tho' both shrapnel and bullets were flying overhead. The men's welcome to me was very warm and hearty and they were most grateful to greet me once more.[60]

Over the following weeks McKenzie was kept busy conducting burial services, including that of his battalion commander.[61] Being exposed to enemy fire, these were usually held at night. Even so it was a dangerous task. He reported having 'had three narrow squeaks': two from bullets, one of which grazed his head, and the other his ear. The third was when he was nearly smothered by an exploding shell.[62]

By the end of a week the 'gleeful happy mood' he reported on the first day of the landings had changed considerably. The Turks launched a major attack on 18 May, and McKenzie was heavily involved in caring for the wounded and burying the dead. The 'Fighting Mac' who 'longed to get at them' was now shocked beyond anything he had previously

---

57 McKenzie, *Diary*, 60.
58 McKernan, *Australian Churches at War*, 50; Reynaud, *Anzac Spirituality*, 246.
59 Gladwin, *Captains of the Soul*, 43.
60 McKenzie, *Diary*, 67.
61 McKenzie, *Diary*, 64.
62 Gladwin, *Captains of the Soul*, 43.

known or imagined. The true horror of war revealed itself to him in a way that changed forever the naïve 'Fighting Mac' of the boxing ring into the veteran padre who, like the Christ he served, was 'a man of sorrows and acquainted with grief'.[63] His diary records his shock:

> I had a very trying duty next day burying our dead. I thought so much of the many sad hearts in Australia when they know of their losses. We laid 28 in one grave all in a row. Monday May 24th was a terrible day. An armistice to bury the dead was granted from 7.30 am to 4.30 pm … It was the most awful sight I have ever witnessed, thousands lay dead along the front, many more than we had anticipated. I read the burial over something like one hundred and twenty men … the Turks had at least 3,500 dead to bury. The smell was unbelievable … I never had such a task and I hope I never shall again. War is indeed 'Hell' & no adequate description can picture its ghastliness.[64]

Walden left Sydney on 24 June aboard the *Ceramic*[65] and arrived at Gallipoli on 19 August. He replaced Robertson, the first Congregationalist chaplain, who had been there with the 6th Light Horse Regiment since 15 May,[66] and in September was diagnosed with pyrexia (fever) and medically evacuated to Egypt, then home to Australia. One of the first dying soldiers Walden ministered to was Chaplain Andrew Gillison – the first Australian chaplain to lose his life – who was dying of wounds received while trying to rescue a wounded soldier.[67] Like Robertson, Walden was only on the peninsula for a short time. He contracted enteritis (extreme diarrhoea) and on 7 September was medically evacuated to Malta, then sent to Florence to recuperate. He returned to Alexandria on 30 November and was posted to the 4th Infantry Brigade.[68] Then, on 5 April 1916 he joined the 5th Australian Division 'for hospital and other duties in Egypt'.[69]

---

63 Isaiah 53:3.
64 McKenzie, *Diary*, 69.
65 Chaplains [alphabetical], AWM 8, 6/6/3, 15.
66 Though a mounted unit, they fought on Gallipoli as normal infantry.
67 Nutt, 'Military Chaplains: For Service of our Soldiers,' 20.
68 Walden G.T. Military Record, NAA B2455, 6-7.
69 Walden G.T. Military Record, NAA B2455, 35.

By the time Walden returned to Egypt the Gallipoli Campaign was coming to an end. Les Carlyon records how in November 1915 the Turks had thrown a note into the Australian lines at Lone Pine, which read: 'We can't advance; you can't advance. What are you going to do?' The answer came on 7 December when the British Cabinet decided to evacuate all troops from the peninsula.[70] Miles and McKenzie both remained with their units until the evacuation took place. McKenzie received orders to evacuate on 15 December.[71] Miles followed him ten days later.[72] Walden, while looking forward to seeing his battalion rather wistfully reflected:

> I am afraid from what I hear, all my effects are now being enjoyed by the Turks. My sleeping bag, my full kit of clothing, rubber boots, mackintosh, camera, field glasses, communion service, surplus stocks for the boys – everything I possessed (except 2 suits of cloths [sic], a helmet, 2 pairs of socks, and a pair of boots), will be doing foreign mission service, diverted from home mission.[73]

Bean described the Australians and New Zealanders who returned from Gallipoli to Egypt as 'a different force from the adventurous body that had left Egypt eight months before'. They were now battle-hardened veterans, and a military force with strongly established traditions.[74] In Egypt they were joined by the newly arrived 8th Brigade and another thirty thousand reinforcements who had arrived too late to join in the fighting.[75]

The following months saw a renewed surge of volunteers in Australia, and more than a hundred thousand enlisted. Despite Lord Kitchener's[76] intention that they become part of an 'army of twelve infantry and two cavalry divisions to defend Egypt', it soon became

---

70 Les Carlyon, *Gallipoli*. (Sydney: Macmillan, 2001), 517.
71 McKenzie, *Diary*, no page recorded (he stopped numbering pages after page 71).
72 Miles F.G Military Record, NAA, B2455, 32.
73 *Australian Christian*, 10 February 1916, 108f.
74 Bean, *Anzac to Amiens*, 183.
75 Bean, *Anzac to Amiens*, 185.
76 The British Secretary of State for War.

clear that they were needed more urgently on the Western Front.[77] The result was the formation of an Anzac Mounted Division from the mounted rifles and light horse brigades,[78] who remained in Egypt to continue the fight against the Turks, and the creation of I and II Anzac Corps,[79] the first units of which sailed for France on 13 March 1916.

Teece was the first of three OPD chaplains (all Congregationalists) to serve in the Sinai campaign.[80] He sailed from Melbourne on 23 November 1915 aboard the *Ceramic*[81] with the 7th Light Horse Regiment, transferring to the 6th Light Horse Regiment on 19 February 1916 while in Egypt. While the four Australian infantry divisions on the Western Front were winning for themselves 'a reputation for reckless valour, their comrades of the Light Horse were also creating a legend in their own right'.[82] The Battle of Romani[83] (3-5 August 1916) established their reputation as elite mounted troops. The 6th Light Horse Regiment was part of the Anzac Mounted Division, commanded by Lieutenant General Chauvel,[84] for whom Romani was 'a personal triumph'.[85] The Adelaide *Chronicle* reported that one other person who distinguished himself at Romani was Chaplain Teece who was awarded the Military Cross for gallantry and devotion to duty, and later was also Mentioned in Despatches. In a situation of great personal danger, he distinguished himself by rescuing a wounded man and then continuing to tend the wounded throughout the battle.[86]

Bean records that the Light Horse had an unwritten law that 'no wounded man should be allowed to fall into enemy hands'.[87] Teece risked his life to ensure that this would be so and won the admiration of his comrades. Not only was he one with the troopers in their hardships

---

77 Peter Firkins, *The Australians in Nine Wars, From Waikato to Long Tan* (Sydney: Pan, 1982), 68.
78 Grey, *A Military History of Australia*, 99.
79 An army corps is a battlefield formation consisting of two or more divisions.
80 This campaign began in 1915 with a Turkish attempt to seize the Suez Canal and ended with the Armistice of Mudros in 1918.
81 Chaplains [alphabetical], AWM 8 6/6/3, 15.
82 Firkins, *The Australians in Nine Wars,* 88.
83 An Egyptian town located 37 kilometres east of the Suez Canal.
84 Bean, *Anzac to Amiens,* 272.
85 Bean, *Anzac to Amiens,* 282.
86 *Chronicle,* 4 November 1916, 41; Teece A.H. Military Record, NAA B2455, 33.
87 Bean, *Anzac to Amiens,* 282.

and danger, but he also epitomised the spirit and ethos of the Australian Light Horse. He finished his deployment later that year, sailing from Suez aboard the hospital ship *Ayrshire* on 21 November and was discharged from the Army on 14 January 1917.[88]

Meanwhile, in France, the Australians discovered that 'war in the Middle East was a poor preparation for the Western Front'.[89] They entered the line in time to participate in the horrendous Battle of the Somme, where in six weeks they lost twenty-eight thousand men, 'the same as for the eight months at Gallipoli'.[90]

The Gallipoli veterans Miles, McKenzie and Walden, were among the chaplains who went with them. Gladwin remarks that the 'massive scale of warfare on the Western Front meant that chaplains undertook a wider variety of duties over a much larger geographical area'.[91] He also mentions that whereas at Gallipoli chaplains often lived 'in the line' with the fighting soldiers, some commanders ordered chaplains to remain in the rear areas away from the fighting. It was a practice that was strongly opposed, and sometimes ignored by many chaplains, who understood 'the difficulty of securing the respect and affection of troops unless they shared similar dangers.[92]

Miles, McKenzie and Walden exemplified these comments. Miles, who was appointed OPD Senior Chaplain on 1 March 1916 and advanced to Chaplain Class 3 (Major),[93] described how during May and June 1916 he was the only Australian chaplain in the Bois Grenier Salient and spent forty-two days there without relief. He reported being under 'severe shell-fire' for 5 days continually and 'had no rest from attending to the wounded'.[94] For his services he was awarded the Distinguished Service Order (DSO). His citation states:

---

88 Teece A.H. Military Record, NAA B2455, 37, 39.
89 Grey, *A Military History of Australia*, 98-100.
90 Les Carlyon, *The Great War* (Sydney: Macmillan, 2006), 244.
91 Gladwin, *Captains of the Soul*, 55.
92 Gladwin, *Captains of the Soul*, 56.
93 On 14 September 1918 he was advanced to Chaplain Class 2 (Lt. Colonel); Miles F.G. Military Record, NAA B2455, 9.
94 Miles F.G. Military Record, NAA B2455, 103; David Harrower, 'Senior Chaplain; Frederick James Miles, DSO, OBE, VD, MID'.

> During Operations in the Fleurbaix Sector, Chaplain Miles did very valuable work among the men in the front-line trenches. When the Battalion went back into Reserve Chaplain Miles remained with the relieving unit as he considered that his place was among the men who were in most danger & enduring the greater discomfort. At Pozieres he was at the Advanced Aid Post where he rendered valuable service in assisting with the wounded & burying the dead. This work being done at considerable personal risk. During the time that the Division has been in the Ypres Salient Chaplain Miles has rendered splendid service.[95]

In 1917 Miles, along with the other Senior Chaplains, was transferred to the AIF's Administrative Headquarters in London[96] from where he coordinated the postings and activities of all OPD chaplains. Except for 'quarterly visits to the Front to see his chaplains',[97] this was the end of his period of battlefield incarnational ministry, which he had helped pioneer and practised so well.

McKenzie continued to live in the front line with the troops, and took part in the great battles at Pozieres, Bullecourt, Mouquet Farm, Polygon Wood and Passchendaele, enduring with the men the terrible winter of 1916–1917. One soldier wrote of him:

> A mile or two beyond the barbed wire entanglements ahead of the main line was what we called the sacrifice line. Each battalion sent a platoon there. No chaplain was supposed to go forward, but McKenzie was back and forth more than any of the men.[98]

Having already been Mentioned in Despatches 'for distinguished and gallant services rendered during the period of ... the Middle East Expeditionary Force',[99] on 3 June 1916 he also was awarded the Military

---

95 Miles F.G. Military Record, NAA B2455, 103; David Harrower, 'Senior Chaplain; Frederick James Miles, DSO, OBE, VD, MID,' accessed 19 November 2021.
96 Miles F.G. Military Record, NAA B2455.
97 Miles F.G. Military Record, NAA B2455, 104.
98 *Sun*, (Sydney), 24 April 1972, 12; David Woodbury, 'Do You Think I'm Afraid to Die with You', *Halleluiah*, Volume 1, Issue 3, Autumn, 2008, 16.
99 Miles F.G. Military Record, NAA B2455, 20.

Cross 'for distinguished service in the Field'.[100] Bean, referring to McKenzie's close personal identification with the troops, described him as 'the most famous chaplain of the 1st AIF'.

Walden, with the newly formed 50th Battalion, arrived in France on 12 June.[101] Like Miles and McKenzie he also chose to live with the troops, sharing their privations and dangers. Writing from "Somewhere in France" he said:

> Am now "billeted" (?) in an oat field, where I eat and sleep. We are only allowed one blanket; this I put on the ground, put my overcoat on, and the rug the Mosman church gave me over me, a bundle of oats for my pillow ... But the ground is like Scrooge's heart, very hard, but sleep comes. But soon something like an earthquake and gunpowder factory explosion, wakes me up. I jump up to find how near the shell has fallen and find it fully 300 yards away, so I lie down to sleep. It is astonishing how used one gets even to shells.[102]

He described his work as 'healthy,' and said he never felt better in his life, although sleeping in his clothes and having no change of underwear made him long for clean sheets and a soft bed.[103] He also spoke of his intense admiration for the soldiers: a feeling that McKernan says was shared by most Australian chaplains, who 'quickly learned to appreciate their real worth' and conceded 'that the men were virtuous and good, apparently independently of formal religion'[104] Walden wrote:

> Our Brigade has been in the hottest corners of the war in France, at Pozieres and Mouquet Farm and Ypres, and we are now at ———where some of the greatest struggles have taken place. In all the fighting and fatigue work our Australian boys have done splendid service ... if Australia could see her sons

---

100 McKenzie W. Military Record, NAA B2455, 21.
101 Walden G.T. Military Record, NAA B2455, 7.
102 *Australian Christian*, 5 October 1916, 599.
103 *Australian Christian*, 5 October 1916, 599.
104 McKernan, *Australian Churches at War*, 62; Reynaud's research, likewise, reveals that despite their larrikin image, most Australian soldiers 'maintained a reasonable standard of ethics', Reynaud, *Anzac Spirituality*, 232; see also Stuart Piggin and Robert Linder, *Attending to the National Soul* (Melbourne: Monash University Publishing, 2020), 63.

now she would say the worst of them is worthy of the highest respect ... It is a constant wonder to me to know how these men who two years ago never handled a rifle, now rank with the world's greatest soldiers.[105]

Walden was advanced to Chaplain Class 3 (Major) on 17 March 1917[106] and continued to serve with the 50th Battalion right through to the end of the war, where they were constantly in action. He was with them during the ferocious German Spring Offensive,[107] when they repulsed the German attack at Dernancourt on 5 April and participated in the legendary re-capture of Villers Bretonneux on 25 April.[108] The editor of the *Australian Christian* wrote of him:

> He has been the faithful friend, comforter, and companion of hundreds of our heroic men. He has taken comforts to them right up to the firing line in Gallipoli and in France; he has cared for them on the battlefield, tended to the sick and dying, reverently buried the dead. He has written hundreds of letters from his sympathetic heart, many of them to loved ones of those he has watched pass into the glory land.[109]

Of the remaining OPD chaplains who followed Miles, McKenzie, Walden and Teece, special attention should be drawn to Benjamin Orames of the Salvation Army, who less than three weeks after his arrival in France, found himself thrown into the maelstrom of the disastrous attack on the German trenches at Fromelles on 19 July 1916, where 'the 5th Division, newly arrived ... suffered 5533 casualties in a single 24-hour period'.[110] His Commanding Officer's report mentioned him specifically:

> In particular I would mention Chaplain Orames' great influence during this battle. The conditions we faced were extremely

---

105 *Australian Christian*, 1 February 1917, 84.
106 Walden G.T. Military Record, NAA B2455, 7.
107 A series of German attacks along the Western Front that created a huge salient in the Allied line and seriously threatened its integrity.
108 Nutt, 'Military Chaplains,' 22.
109 *Australian Christian*, 2 August 1917, 453.
110 Grey, *A Military History of Australia*, 102.

severe, with terrible weather conditions and enemy fire. He worked throughout these conditions, quietly going among the men. And I am certain enabled a great number who would have otherwise given in, to stick it out.[111]

George Cuttriss, who twice narrowly escaped death, including having 'been blown a dozen feet by a mortar shell',[112] was also highly regarded by his commanding officer, who recognised the value of his identification with the troops and described his work as:

> Satisfactory in every way ... not only from a religious standpoint, but also from a man standpoint, and appears to be working in the hearts of the men in a manner which can only bring the best results, from a disciplinary point of view. He is particularly interested in all that appertains to the men's welfare.[113]

Local commanders were not the only ones to recognise the incalculable contribution to morale that emanated from the incarnational ministry of chaplains like these. The importance of having such people in forward areas was eventually recognised even by Sir Douglas Haig,[114] who in 1916 observed:

> That the troops are in such splendid heart and moral[e], and fight without 'counting the cost', is largely attributable to our chaplains, who have so successfully made our men realise what we are fighting for, and the justice of our cause.[115]

These sentiments were shared by the most senior Australian commanders and were of enormous encouragement to OPD chaplains anxious to prove their worth alongside their colleagues of the major churches. Miles, in a letter to the *Australian Christian*, spoke of the enthusiastic support given to him as Senior Chaplain by the highest levels of AIF

---

111 David Orames, 'The Life of Commissioner Benjamin Orames,' accessed 1 December 2021, https://www.sa-gong-history.com/Benjamin/index.htm.
112 Frederick James Miles, *War Diary, Senior Chaplain, Other Protestant Denominations, Headquarters AIF, London, August 1914–October 1917*, Canberra: Australian War Memorial 4, 6/4/1 Part 1, November 1916.
113 Nutt, 'Military Chaplains,' 23.
114 The British Commander-in-Chief on the Western Front from late 1915 to the end of the war.
115 Gladwin, *Captains of the Soul*, 56.

command, including Birdwood and Monash, indicating their appreciation of the role that OPD chaplains played in support of the troops and their contribution to morale. He recorded that his chaplains 'have received at the hands of all responsible officers, every courtesy and consideration,' and that he himself had 'never been better treated' than he had by the senior officers of the AIF.[116]

Ardent Evangelicalism

The second characteristic of this emerging ethos is ardent evangelicalism. Gladwin refers to Robert Linder's calculation 'that of the four hundred and fourteen clergy who served as chaplains in the Great War, at least one hundred and ninety-five were likely to have been Evangelicals'.[117] This was certainly true of most OPD chaplains. References to 'enquirers', 'decisions', 'reaping precious souls', and 'enlisting them as soldiers of Christ,' along with other expressions of the evangelical terminology common at that time frequently appear in their letters and reports.

David Bettington defines Evangelicalism as possessing four main qualities: '*conversionism,* the belief that lives need to be changed; *activism*, the expression of the gospel in effort; *Biblicism*, a particular regard for the Bible; and what may be called *crucicentrism*, a stress on the sacrifice of Christ on the cross. Together they form a quadrilateral of priorities that is the basis of Evangelicalism'.[118]

Evangelicalism was – and still is – overwhelmingly the theological orientation of the Baptists, Churches of Christ and the Salvation Army. Ian Rennie traces evangelical theology back to the earliest days of the Christian era and the conviction that 'the Bible is the truthful revelation of God and through it the life-giving voice of God speaks'. He also relates it to the evangelical awakenings of the eighteenth century in which the 'nature of saving faith, or conversion,' was continually to the

---

116 *Australian Christian,* 10 July 1919, 399.
117 Gladwin, *Captains of the Soul,* 66; Robert Linder, *The Long Tragedy: Australian evangelical Christians and the Great War, 1914–1918* (Adelaide: Open Book, 2000), 22-23, 125.
118 D.W. Bettington, *Evangelicalism in Modern Britain: A History From The 1730s To The 1980s* (London: Routledge, 1989), 2f.

fore.¹¹⁹ The prospect of redemption and eternal life was a powerful message to men who faced death daily.

The only possible exceptions to an OPD evangelical consensus are the Congregationalists. Hugh Jackson, in his study of Congregationalism in South-Eastern Australia, refers to an address delivered at the Colonial Missionary Society meeting on 14 May 1880 in which it was affirmed that: 'no people in the world preach the Gospel in greater purity than our [Congregational] ministers in Victoria'.[120] However, he reports that in the years leading up to the Great War most Congregational ministers had embraced a more liberal theology, despite there still remaining many who 'were conservative and sought to shield their people [from liberalism]'.[121] Teece appears to have been one of them. A newspaper report said of him: 'Mr Teece preaches a broad evangelical gospel avoiding inflated oratory and cant phrases'.[122]

Whether the other Congregationalist chaplains shared his evangelicalism is difficult to judge. Two of them, Frank Dowling and Reginald Turner, were later involved in the Australian work of the London Missionary Society, and Harold Perkins served for many years in Samoa,[123] which still embraces a very evangelical form of Christianity. This may indicate they also were evangelicals. On the other hand, Dowling was also a lecturer at the United Faculty of the Presbyterian, Methodist and Congregational Churches, which could suggest his was a more liberal theology.

McKernan reports that the only official duty chaplains had on the troopships was 'to conduct a church parade each Sunday'.[124] But many chaplains found the voyage to war a fruitful opportunity for ministry. Miles said of it:

> It has been an immense privilege to serve these splendid fellows ... the work was difficult, as we were tightly packed, and

---

119 Ian S. Rennie, 'Evangelical Theology,' in *New Dictionary of Theology*, ed. Sinclair B. Ferguson and David F. Wright (Leicester: IVP, 1988), 239.
120 Hugh Rutherford Jackson, Aspects of Congregationalism in South-Eastern Australia, circa 1880–1930', Unpublished Ph.D. dissertation, ANU, 1978, 169.
121 Jackson, *Aspects of Congregationalism*, 182.
122 *The Mail*, 17 April 1915, 9.
123 'Australian Chaplains in WW1', accessed 26 January 2022. https://ww1chaplains.weebly.com.
124 McKernan, *Australian Churches at War*, 47.

there were no rooms where meetings could be held ... We had church parade every Lord's Day morning, and a voluntary service at night ... On week nights I ran several illustrated lectures and concerts ... Every day I visited the sick ... At certain hours I was in my cabin to meet enquirers ... I had the joy of seeing several fine decisions for Christ.[125]

McKenzie's diary records how having departed Sydney he began a daily routine of organising official prayers. While at anchor at Albany he conducted the burial service for 'a fine sweet-spirited young man of 21 years', three well-attended Sunday morning services, an afternoon bible class, a short hospital service and a stirring evangelical service at 8 pm'.[126] It was a pattern that he and Miles continued to observe throughout the voyage.

Crossing the Indian Ocean was momentous also in that it included Australia's first naval victory when the *Sydney* left the convoy to seek out and destroy the German commerce raider *Emden*. Later, having left Aden, they received a signal ordering them to disembark in Egypt and complete their training there. They landed at Alexandria, travelled by train to Cairo, then marched out to the edge of the Libyan Desert, where, at the Maadi and Mena Camps, the whole division came together for the first time, and 'the task of welding the division into a single instrument of war' began.[127]

The popular larrikin image of those original Anzacs is evident in a description of them by Fr William Devine, one of the Roman Catholic chaplains:

They were not a kid-glove lot of men, and required something firmer than kid-glove handling. Those of them who drank, drank deep and were noisy in their cups and strong in their language. Most of them were at that time ill-trained soldiers, or not trained at all ... All in all they were good fellows, with a manly simplicity of character.[128]

---

125 *Australian Baptist*, 25 January 1916, 11.
126 McKenzie, *Diary*, 4-6.
127 Bean, *Anzac to Amiens*, 64-65.
128 William Devine, *Story of a Battalion* (Melbourne: Melville and Mullen, 1919), 116-17, cited in Johnstone, *The Cross of Anzac*, 34.

Their need for serious military training is emphasised by Jeffrey Grey, who concludes that 'for all the courage of its soldiers, the 1st Division was probably the worst-trained formation ever sent from Australian shores'.[129] Their need for spiritual guidance and pastoral support was also very evident to their chaplains, especially the evangelicals.

Miles reported that he had had 'a great time' at Mena Camp. He helped erect the first YMCA hut and established a Christian Endeavour society with over one hundred members, describing their meetings as 'the best that I have ever attended'.[130] He also 'preached at [the] Cairo Methodist Church, Cairo YMCA, the American Mission at Heliopolis, and the Territorials YMCA':[131]

> The last was packed with troops, and large numbers attended the other services. There were several manifest results. To God's glory, be it stated, that for many Sunday evenings in succession I had the joy of reaping many precious souls at our own YMCA service. Many of these are standing firm to this day.[132]

Walden's fervent evangelicalism is also evident in his letters home. While ministering at the Gazireh Palace Hospital in Egypt in early 1916[133] he listed various activities he undertook to support the troops, concluding with: 'seeking to enlist them as soldiers of Christ'.[134] Later that year, amidst the squalor and misery of the Western Front, another letter reveals both his pastoral heart and his evangelistic passion. He said that his saddest work was burying those 'brave boys who have given their lives for the Empire' and now 'rest under the shade of some lovely elm trees in one of France's beauty spots; "far from home and kindred"' and that he hoped they were 'asleep in Jesus'.[135]

Throughout the war an evangelistic imperative – the salvation of souls – dominated the ministry of most OPD chaplains, as is evident in Miles' report on the activities of his chaplains in August 1918. One

---

129 Grey, *A Military History of Australia*, 93.
130 *Australian Baptist*, 25 January 1916, 11.
131 The British Territorial Army, which was the equivalent of the Australian CMF.
132 *Australian Baptist*, 25 January 1916, 11.
133 Walden G.T. Military Record, NAA B2455, 33.
134 *Australian Christian*, 23 March 1916, 184; Nutt, 'Military Chaplains,' 21.
135 *Australian Christian*, 5 October 1916, 599.

chaplain announced that fifty-seven men had 'professed their faith in the Lord Jesus Christ', and signed a pledge to remain faithful, whatever the cost. Another described how it had been his 'extreme joy to point 316 to Christ during that month's services'.[136] Regardless of ways in which the war might have affected other aspects of their theology, for most OPD chaplains ardent evangelicalism remained a constant characteristic.

Trans-Denominationalism

Trans-denominationalism – or practical ecumenism as it was later known – is another feature of the emerging ethos. It was the very thing the Navy originally wanted for its chaplaincy branch but failed to receive because of ecclesiastical parochialism. Despite the Chaplain Generals' policy of denominational chaplaincy and the sectarianism endemic in Australian society, the very nature of shipboard life and battlefield deployments required chaplains to be positioned where they could be of most value to all. OPD chaplains responded readily to this, and easily embraced the trans-denominationalism that developed on active service.

Teece was especially outspoken in his advocacy of such ecumenism. Even before his deployment to the Sinai he bemoaned the fact that though 'all parties in the State had sunk their differences of opinion and joined together', the same could not be said of the churches, especially in Australia, where the lack of unity was a 'great drawback' to their welfare.[137]

Miles frequently referred to the unity among the chaplains with whom he worked.[138] His early reports from Egypt reveal the emergence of a trans-denominational outlook that was to largely dominate chaplaincy on active service. He referred to being chosen by his 'brethren at a chaplains' meeting to conduct Sunday afternoon services for over 500 VD men, in a specially isolated camp':[139] the same 'primitive and

---

136 *Australian Baptist*, 24 December 1918, 2; Petras, *Australian Baptists and World War I*, 25.
137 *The Mail*, 17 April 1915, 9.
138 *Australian Baptist*, 16 May 1916, 2.
139 *Australian Baptist*, 25 January 1916, 11.

humiliating compound' where McKenzie led an assault to 'tear down the barbed wire entanglements to which the men particularly objected'.[140] Miles even allowed the trans-denominational reality of military chaplaincy to coexist with his convictions concerning believers' baptism: a fundamental tenet of both Baptists and Churches of Christ. In a letter to the *Australian Baptist,* he wrote:

> To me confirmation is a church invention ... substituting 'that ordinance of man,' infant baptism, for believers' baptism, as ordained by Christ. Yet on a recent Lord's Day I found myself at three camps on the front line, asking those who wished to be confirmed to see me ... The following day I rode in to Serapeum ... and handed a list of their names ... to a C of E chaplain.[141]

It was at Gallipoli that the seed of trans-denominationalism came into full bloom. Gladwin speaks of the 'close friendships and professional respect [that] developed among padres of all denominations,'[142] as they organised themselves to provide the widest coverage. Miles gladly took his turn with others, whom he again referred to as 'my brethren', in visiting the sick and wounded,[143] and McKenzie during the three weeks following the Australian assault that began on 6 August, reported that he 'was the only Protestant chaplain in the Brigade',[144] working closely with his Anglican and Catholic colleagues.

It was a pattern that increasingly became the norm throughout the war. Walden, during his period of duty in Egypt, also found the 'denominational boundaries of chaplaincy were becoming blurred'. At the Gazireh Palace Hospital he ministered to men from 'thirty-two battalions and many brigades of light horse, conducting services and visiting the sick' regardless of denomination.[145] On 2 February 1916 he wrote:

---

140 McKernan, *Australian Churches at War,* 49.
141 *Australian Baptist,* 16 May 1916, 2.
142 Gladwin, *Captains of the Soul,* 67.
143 *Australian Baptist,* 25 January 1916, 11.
144 McKenzie, *Diary,* unnumbered but 79 (estimated).
145 Nutt, 'Military Chaplains,' 21.

> My work at the hospital is to conduct services on Sundays and visit the sick during the week. I take all who are not R.C., Church of England, or Presbyterians. I have Jews, Theosophists, Brethren, Baptists, Congregationalists, Church of Christ, Lutheran, Dutch Reformed Church; and since Col. (Chaplain) Nye left, Wesleyan and Methodist.[146]

The harsh realities of war caused many chaplains to reflect on the Church at home's failure to fulfil Christ's prayer that 'all of them may be one, ... so that the world may believe'.[147] Cuttriss, comparing his experience in France with the situation at home, wrote:

> The war is an indictment against divided Christendom ... Men on active service have grown indifferent, not to Christ and his church, but to human creeds ... When I left Australia, some rumour led me to believe that the men were drifting away from God, and the churches were drawing nearer to Christian union. From observation and experience I unhesitatingly affirm that such is not so. The men are getting nearer God. The churches are drifting apart ...[148]

This trans-denominationalism was a shared ministry that sprouted in the first Anzac convoy, where chaplains from different denominations would preach at united church parades on a rotating basis,[149] as well as at the informal meetings that Miles and McKenzie described. It continued to grow in the training camps of Egypt, blossoming at Gallipoli where the nature of the campaign required individual chaplains to serve all the troops in certain areas regardless of denomination. By the time the AIF reached the Western Front it had become firmly established, and the yardstick was a 'chaplain's willingness to participate in a united service'.[150]

The relative ease with which OPD chaplains adapted to this early expression of ecumenism can be partly explained by factors inherent in

---

146   *Australian Christian*, 16 March 1916, 184.
147   John 17:21, New International Version.
148   *Australian Christian*, 5 July 1917, 296.
149   McKernan, *Australian Churches at War*, 47.
150   McKernan, *Australian Churches at War*, 138.

their denominational backgrounds. Ecumenism had become increasingly important for Congregationalists in the years preceding the war.[151] Jackson claims that it was present from the very beginnings of Australian Congregationalism:

> The Australian Congregational churches were influenced by the non-sectarian ethos from their inception. The first recorded Congregational church, established in 1810, was formed by agents of the London Missionary Society, which epitomized the intensely ecumenical spirit characteristic of English Protestantism when the Evangelical Revival was at its height.[152]

For the Churches of Christ, the years 1875–1910 were a time when their central focus shifted from an emphasis on restoration to one of unity among the denominations, evident in a greater willingness to engage in dialogue with other Protestant churches and to join with them in efforts to stem the tide of social evils.[153] In both New South Wales and Western Australia committees were set up to explore the possibility of union between the Baptists and the Churches of Christ.[154]

The Baptists, though less inclined towards organic union, were significantly involved in interdenominational activities. Ian Breward notes that their 'associational form of unity ... influenced from Britain and the United States ... has kept them out of much mainstream ecumenism, but ... has been important in para-church organisations'.[155] By the start of World War I, support for such interdenominational mission work was well established in Australian evangelical churches, especially Baptist churches.[156]

The Salvation Army, whose founder William Booth, as previously noted, intended it to be a mission to the poor rather than a church, had in reality become a church by the start of the war.[157] Vivian Green

---

151 Ian Breward, *A History of the Australian Churches* (Sydney: Allen & Unwin, 1993), 99-100.
152 Jackson, *Aspects of Congregationalism*, 30.
153 Graeme Chapman, *One Lord, One Faith, One Baptism; A History of Churches of Christ in Australia,* Melbourne: Vital, 1979, 104-106.
154 Chapman, *One Lord, One Faith, One Baptism,* 107.
155 Breward, *History of the Australian Churches,* 220.
156 Stephen Neill, *A History of Christian Missions* (Melbourne: Penguin, 1964), 335-336.
157 Williston Walker, *A History of the Christian Church* (Edinburgh: T&T Clark, 1959), 501.

describes it as 'a new religious force ... [that] met a religious need that the more conventional churches were unable to supply ... [and] whose social activities gave it a moral authority'.[158] However, its essential ethos, born in the slums of 'Darkest England',[159] ensured that it would be an outward looking body, ready to minister to all, irrespective of race, colour and creed, and would work in association with other religious bodies. It was, therefore, quite natural for its chaplains, along with those from the Congregational, Baptist and Churches of Christ, to work in a trans-denominational environment.

The most noticeable exception to this spirit of trans-denominationalism was in the divide that separated Roman Catholics from all other Christian denominations. Gladwin refers to the 'widespread acceptance of the separate identity of Roman Catholics, which had long been maintained and reinforced in Australia through a separate educational system and subculture'.[160] The theological differences between the various Protestant groups, though often contentious, were much more easily accommodated than those that separated them from the Roman Catholics. This was especially true of evangelicals, whose acceptance of the Bible as the supreme revelation of God's will to humankind could not tolerate Catholicism's dogmas of Papal infallibility and the authority of the church. Nor could they reconcile the gospel of grace received by faith with Roman Catholic teachings of the mass and that salvation is only found through the (Roman) Catholic Church.[161] Even so, at a personal level, OPD chaplains often enjoyed friendships and mutual admiration with their Catholic colleagues, as is evident in chaplain Dowling's (Congregational) letter to the sister of Roman Catholic chaplain Edward Sykes, who was killed in November 1918. Dowling spoke of him as:

> A most cultivated man, a distinguished minister of his Church, a man quite out of the common ... there in the disorder and

---

158 Vivian Green, *A New History of Christianity* (Stroud: Sutton, 1996), 303.
159 *In Darkest England and the Way Out*, published in 1890, was William Booth's detailed plan for ending unemployment and overcoming poverty.
160 Gladwin, *Captains of the Soul*, 68.
161 J.E. Colwell, 'Roman Catholic Theology,' in *New Dictionary of Theology*, ed. Sinclair B. Ferguson and David F. Wright (Leicester: IVP, 1988), 597-98.

distress and agony of the time, mingling with his direct spiritual counsel the very lowly service of handing out cocoa and cigarettes ... lending a hand in any way he could ... invested with the true essential dignity of his calling.[162]

Servant Leadership

Servant leadership is important to this emerging ethos. McKernan noted that the battlefield is a strange place to find Christian ministers, who are dedicated to the message of peace, 'But, if men are to die or suffer, tradition decrees that their ministers should be with them'.[163]

Though they were commissioned officers and entitled to the privileges of rank, OPD chaplains gave of themselves unstintingly in serving the men in the trenches. Gladwin explains how chaplains in France played a key role in 'providing for soldiers' social, mental and physical welfare', setting up clubs behind the front line where soldiers could spend their recreation time.[164] It was a new form of chaplaincy ministry that was first pioneered in the training camps of Egypt, especially by Miles and McKenzie. Bean makes special mention of McKenzie's (and other chaplains') efforts to organise amusements for the men, making up for the deficiencies of the Divisional staff who failed to give serious attention to the recreation of the troops.[165] This noticeable lack was one of the reasons for the number of men confined to the detested VD camp. STDs are a perennial concern for military authorities, but in Cairo they did little to prevent their spread. Sergeant Archie Barwick said of the Cairo brothels:

> Once inside these dens unless you have a very strong will, you are done for. They are places of the vilest description where the inmates would sell their souls for sixpence.[166]

---

162 Gladwin, *Captains of the Soul*, 68.
163 McKernan, *Padre*, x.
164 Gladwin, *Captains of the Soul*, 70.
165 Bean, *Anzac to Amiens*, 66.
166 Archie Barwick, *Diary*, AWM F940.26093 B296d, 9, in Peter Fitzsimons, *Gallipoli* (Sydney: Random House, 2014), 121.

Commenting on the absence of authority in those places, Peter Fitzsimons observes that McKenzie, whom he describes as 'a beloved Salvation Army chaplain' was the 'one exception' who would regularly visit the "Wazza" to drag drunken Australians out of the brothels and put them on the tram back to camp, before they can disgrace themselves'.[167] One of the strategies he employed to keep young soldiers out of the vice dens was to organise boxing bouts, a sport in which he himself excelled. McKernan comments that 'his long reach, jarring uppercuts and dangerous half-hooks left some of the AIF's best fighters dazed'.[168] The journalist Keith Murdoch said of him: 'Those with whom he was associated soon appreciated him as a true friend, a genuine "cobber", and a wise guide in the midst of manifold temptations'.[169] The greatest tribute probably came from Bean who, years later in a radio broadcast, said:

> In 1914, when our troops reached Mena Camp … pressure of training gave the staff no chance to provide amenities to counter the deadly attractions of that sink of iniquity … Several chaplains (but Mac the foremost) set to organise amusement … No matter how low a man came, Mac was his friend – men felt he would stand by you when you were past all other help; and this gave him his fame.[170]

Servant leadership was especially evident in the mobile canteens that many chaplains operated, providing troops with inexpensive items that made the misery of the trenches more bearable. McKenzie, true to his background as a Salvation Army officer, supported the soldiers with coffee stalls and other 'comforts' like soap, tobacco and fruitcakes.[171] But no one did this better than Walden, 'the Canteen King' as he became known.[172] He described his *modus operandi* in a letter to the *Australian Christian*:

---

167 Peter Fitzsimons, *Gallipoli*, (Sydney: Random House, 2014), 122.
168 McKernan, *Padre*, 3.
169 Woodbury, 'Do You Think I'm Afraid to Die with You', 14.
170 Charles Bean, 'Fighting Mac', *News Digest* broadcast, 28 July 1947.
171 'McKenzie, William,' Australian Chaplains in WWI', ww1.chaplains.weebly.com.
172 Miles, *War Diary*, July-December 1916.

> I get the loan of a cart, drive to some big canteen, and spend every centime I possess, come home, open up my goods, and begin to sell. I have cigarettes, tobacco, sardines, tinned fruits, lobsters, salmon, chocolate, matches, chewing gum, lollies, post cards, paper, pencils, envelopes, etc … I sell [to] the boys at cost price.

He then went on to describe the problems he has with the temperamental mule that assists him in this mission of mercy:

> The mule they give me … is a very pious-looking animal … but he is very conservative, and when a motor cycle or motor car comes along he darts for the field, and not even the restraint of 'Holy Church' can hold him. After he has dragged me over the standing crop or the ploughed field he resumes his innocent ways and we jog along until the next sign of modern life appears.[173]

In another letter he described how his battalion had been on the move for three months, having been in four different places in the front line, and how he ministered to the men by operating four canteens and some coffee stalls. He remarked that:

> Unlike the Apostle Peter … I have tried to do the 'serving of tables' and have not left the word of God, as for some Sundays have walked ten miles and conducted as many as three church parades.[174]

Walden was a prolific letter writer. While serving at the Gazireh Palace Hospital in Cairo, in addition to comforting the dying and burying the dead, he wrote letters for soldiers unable to do so, did their banking, talked of home, looked at photographs, put them in touch with others from the same hometown, replaced lost Testaments and distributed Red Cross parcels.[175] His letter writing continued when he was posted to the Western Front, and he records that by 5 October 1916, in addition to personal letters to his family, he had written a further eight hundred and

---

173 *Australian Christian,* 5 October 1916, 599.
174 *Australian Christian,* 1 February 1917, 84.
175 *Australian Christian,* 16 March 1916, 184.

fifty-five. Urging people back home to also write to soldiers at the front, he described the effect such letters had:

> Letter day with the soldiers is a big event in the soldiers' lives ... You can see men sitting about the camp reading their letters with smiles, as some good news is read, sometimes with sad faces as bad news is read, but smiles predominate. I am sure that no more spiritual uplift comes into the lives of the soldiers than this bit of home life ... There is so much in this life to harden men and tempt them from spiritual wholeness, but when messages come from mother and father, children and sisters and brothers, and the dear girl sweethearts, it speaks to them of God and home and purity and love; and I can notice quite a different tone among all after 'letter day'.[176]

Another dimension to the concept of servant leadership was provided by Henry Procter (Churches of Christ), whose rescue of a young woman has already been mentioned. It came through his skills as an inventor and designer. In February 1918 he was sent on leave to England,[177] where he helped design an apparatus for quickly withdrawing iron tent stakes when circumstances necessitated the rapid relocation of a facility. He also designed appliances to aid surgeons, including a retractor 'for use in severe chest cases which obviates the necessity of cutting one or more ribs'.[178]

Continuity of Service

Continuity of service was essential to this ethos. Gladwin mentions that the minimum period of service for chaplains was twelve months and that many served for a year or less. The reason usually given for this was that longer deployments would adversely affect the health of the church at home.[179] But early return was not true of the OPD 'continuous service' chaplains, whose average length of service, including those who were appointed in the closing stages of the war, was twenty-seven months.

176 *Australian Christian,* 5 October 1916, 599.
177 Proctor H.A. Military Record, NAA B2455, 12.
178 Nutt, 'Military Chaplains,' 26-27.
179 Gladwin, *Captains of the Soul,* 74.

The only indication of a weakening of this consensus came as a result of a letter sent by Chaplain E. Davies, the OPD Senior Chaplain at 3rd Military District Headquarters in Melbourne, to fellow Congregationalists serving with the AIF. One such, dated 15 October 1917, was sent to Harold Perkins, who had only been at the Front since 5 October. Davies urged him to resign his chaplaincy and return to his parish, which was struggling.

Davies insisted that there was no suggestion that Perkins was being recalled, and that his work was deeply valued. He reported that Headquarters (presumably 3rd Military District HQ) had assured him that 'there was no reason why chaplains might not be relieved after twelve months service, that the other denominations were already doing so'. Davies argued that this would then provide opportunities for 'other fit and enthusiastic young men' to serve, and 'derive the benefit that comes from the experience'. He emphasised that there was a limit to how long the churches at home could manage without their minister being present; and that the effects of the war, 'both present and future' required the presence of 'as many qualified and authoritative men as possible to meet the new conditions'.[180]

Perkins waited six months before tendering his resignation to Miles, who reluctantly recommended that it be accepted, noting that Perkins was responding to pressure from his denomination. He added: 'In my judgement he is just reaching full efficiency ... It is deeply deplorable that such a fine fellow should be withdrawn just when he is beginning to be most useful'. He recommended that in future no chaplain should be sent to the Front unless he was prepared to stay longer than twelve months and stressed that he regarded 'long service as essential to success', noting that this was the second time that a Congregational chaplain had resigned having received such a letter from Davies.[181] Perkins' resignation was approved by AIF Headquarters, along with a request that the Secretary of Defence be made aware of the letter that Davies had sent, and that he be told not to do it again'.[182]

---

180 Perkins H.S. Military Record, NAA B2455, 31.
181 The other was Chaplain Gunson; a third Congregationalist, Chaplain Dowling, resigned in October 1918, also citing problems in his parish.
182 Perkins H.S. Military Record, NAA B2455, 31-33.

Non-Coherent No More

Between them, the OPD chaplains provided the equivalent of fifty years' service. At least three of them were wounded and one was gassed. Some of them, especially older men like McKenzie, returned home with their health affected by the rigours of months living in the open. Whether any suffered post-traumatic stress is not known, but it is likely that some did. Although not recognised as a medical illness until 1980, post-traumatic stress is not a modern phenomenon and was then commonly known as shell shock. Knowing, as we now do, that up to thirty percent of recent veterans developed PTSD as a result of their experiences, it is certain that the intensity of the fighting during World War I must have affected those chaplains too.

At the start of the war, they were to a large extent an unknown quantity, viewed with suspicion by the Chaplains General and occasionally by other chaplains who questioned the validity of their ordination and considered them unqualified to minister to any other than soldiers of their own denominations. Miles, for example, complained of a recently arrived Anglican who would not accept the validity of his ordination and insisted on referring to him as a layman. Miles was clearly offended by this but asked for prayers that in his response he would be 'Christian and courteous'.[183] Fortunately, experiences like this seem to have been relatively rare and, even if such feelings were present, most chaplains were discreet enough to keep it to themselves.[184]

But any doubts about their worth as chaplains had, by the war's end, vanished for good. OPD chaplains by their lives and actions won the respect of fellow chaplains and soldiers alike and proved themselves to be among the most effective chaplains of the war. Like most AIF chaplains, they went to war ill-prepared for what lay ahead, usually embarking within a month of their appointment and with no specialised chaplaincy training. They learned their trade by experience, beginning with the long voyages aboard the troopships where hundreds

---

183 *Australian Baptist,* 16 May 1916, 2.
184 Roman Catholic and High Anglican chaplains held strong beliefs about apostolic succession, which may be defined as the uninterrupted transmission of spiritual authority from the Apostles through successive popes and bishops.

of men were cramped together with little to do, organising lectures, concerts and singsongs, as well as worship services and Bible studies. Then, in the training camps of Egypt, in addition to their spiritual duties, they made up for the deficiencies of the divisional staff by organising recreational activities. In France they added to these by providing canteens where soldiers could purchase 'comforts' that made life more bearable. Ignoring official policies that they should remain behind the lines, many of them lived in forward areas and shared the same dangers as the soldiers. And even though many AIF chaplains only served the minimum term of twelve months – and some less – the average deployment of OPD chaplains was twenty-seven months.

As has already been noted, they were generally held in high esteem by their commanders.[185] One example is the letter that Sir John Monash, the Australian Corps Commander, sent to Cuttriss referring particularly to his work as the Divisional Burials Officer and his publishing of an official history of the Australian 3rd Division.[186] Among the cemeteries he established was the famous one at Villers-Bretonneux. It was a dangerous and emotionally traumatising task, which he described in a letter to Miles as 'the least to be desired of all Military Service'.[187] Monash wrote:

> I take this opportunity of extending to you my very best thanks for the splendid service which you have rendered in so many different capacities, in all of which you have been quite indefatigable, and of the greatest service to me and the 3rd Australian Division ... You have also for a long time carried out the difficult and dangerous duties of Burials Officer for the Division ... In addition, the 3rd Division owes to you much for useful work in the historic and literary field, and the works which you have published both in prose and in verse, have reflected renown upon the Division as upon yourself ... You have, at all times and under all circumstances displayed the

---

185 Reynaud, *Anzac Spirituality*, 261-62
186 George Percival Cuttriss, *Over the Top with the Australian 3rd Division* (London: Charles H. Kelly, 1918).
187 Nutt, 'Military Chaplains', 23-24.

very best qualities which have made the Australian soldier famous throughout the world.[188]

Perhaps one of the least known indicators of OPD chaplains' effectiveness is that of the seventy-two honours that were awarded to AIF chaplains, eight – or 11.1 percent – were to OPD chaplains, who only represented 6.5 percent of the four hundred and fourteen chaplains who served.[189] They had earned the respect not only of the soldiers and their commanders, but also of their fellow chaplains.

Under normal conditions it takes time for a freshly hewn stone to weather. But when that stone is constantly exposed to the violence of storm and tempest, the weathering process is much faster. At the beginning of the Great War the chaplains of the Other Protestant Denominations were an unknown and largely untested commodity, treated with a measure of disdain by the Chaplains General. But by the war's end they had proved themselves equal to the best, and in some cases better than most. The once-rejected stone was now fully weathered and permanently in place.

---

188 Miles, *War Diary*, Appendix 5; Nutt, 'Military Chaplains', 25.
189 'Chaplains, Australian Army, First World War', Australian War Memorial.

CHAPTER FOUR

Between the Wars: The Stone in Storage

ON 25 APRIL 1919 – the first Anzac Day following the Great War – Australia was in the midst of the Spanish Influenza Pandemic that swept the world and killed six times as many people as died in the war. The march through the streets of Sydney had been cancelled, and those who gathered in the Sydney Domain to commemorate the occasion were required to wear face masks and stand a metre away from each other. Pride and sorrow mingled as people reflected on how one of the world's newest nations had seen nearly half its eligible male population volunteer to serve, and do so with unsurpassed valour, but at terrible cost. More than half died or became casualties and many thousands more were scarred in mind and spirit by what they had experienced.[1]

In addition to pride and sorrow, there was also a sense of bewilderment. Eighteen years earlier the world had greeted the twentieth century with unbounded optimism. Geoffrey Blainey described it as a 'flaming sunrise' and that 'more was expected of the century than any other'.[2] This was to be the golden age. Universal education and enlightened thinking had finally triumphed over the superstition and

---

1 Bill Gammage, *The Broken Years, Australian Soldiers in the Great War* (Melbourne: Penguin, 1982), 283.
2 Geoffrey Blainey, *A Short History of the 20th Century,* (Melbourne: Viking, 2005), 5.

brutality of the past. Yet, after little more than a decade, that same world descended into an orgy of unimagined savagery,³ until finally, bankrupt and exhausted, it tried to make sense of how it could have happened, while naïvely vowing never to do it again.⁴ Piggin and Linder were correct in asserting that the war had 'threatened the fundamentals both of Western civilisation and of Christian faith',⁵ as was Patrick Porter's assessment that it 'refuted secular thinkers who had believed in humanity's continuous enlightened and rational ascent.'⁶

The *Sydney Morning Herald* reported that the service in the Domain was 'of the simplest character, and was largely attended by womenfolk, many of them in mourning'. The address was given by Chaplain 'Fighting Mac' McKenzie who predicted: 'As the years come and go the day [Anzac Day] will gather added glory and lustre, and we will cherish it more deeply because of the great national and spiritual ideals for which it stands'.⁷

It is of particular significance that McKenzie, whom the *Sydney Morning Herald* described as 'the popular padre "Fighting Mac"',⁸ a chaplain from the smallest member of the Other Protestant Denominations, should have been chosen to deliver the address on this most historic occasion. It would have been inconceivable four years earlier for anyone less than an archbishop – or perhaps a very senior representative of one of the major churches – to have done so. It speaks eloquently of how the once insignificant Other Protestant Denominations had proved their worth in popular perception. In Melbourne, a similar service was held in the Melbourne Cricket Ground. *The Age* described it as:

> Impressive and solemn ... A thanksgiving for the successes attending the battles of the Empire's soldiers against the mighty forces arrayed against them, intermingled with expressions of

---

3   David Shermer, 'World War I,' in *Wars of the 20th Century*, ed. S.L. Mayer (Secaucus, N.J: Derbibooks, 1975), 239.
4   A.J.P. Taylor, 'The War in Perspective,' in *History of the First World War*, Volume 8, ed. Peter Young (London: Purnell, 1971), 3533.
5   Stuart Piggin and Robert Linder, *Attending to the National Soul* (Melbourne: Monash University Publishing, 2020), 91.
6   Patrick Porter, 'The Sacred Service: Australian Military Chaplains in the Great War,' *War and Society*, Volume 20, (2002), 23.
7   *Sydney Morning Herald*, 26 April 1919, 17.
8   *Sydney Morning Herald*, 26 April 1919, 17

the deepest sympathy with bereaved parents and relatives and the everlasting remembrance of the fallen heroes.[9]

Historians trying to explain the importance of Anzac to Australians coined the phrase the 'Anzac Legend'. Charles Bean declared that it stood for reckless valour in a good cause, for enterprise, resourcefulness, fidelity, comradeship, and endurance that will never own defeat.[10] Manning Clark quotes the poet David McKee Wright who described Anzac Day as a 'march through the hearts of men ... an occasion to "hear the trumpet call" to a nobler life'.[11] Clark also recognised in it the emergence of a new and secular religion that was to dominate Australian consciousness for the next hundred years:

> Australians now had a faith, but what that faith was no one could say ... the poets had their shot at putting it into memorable words ... using the language of religion to describe a secular experience. Anzac Day was becoming a secular religion.[12]

It is a view that has been challenged in more recent times. Tom Frame reduces it to 'devotion to the nation state as the highest embodiment of human virtue' and criticises its liturgies as 'unintentionally belittling conventional religion'.[13] Similarly, Marilyn Lake and Henry Reynolds argue that the transformation of Anzac Day into a sacred myth has overshadowed the 'rich and diverse history of nation making and distorted the history of Gallipoli'.[14] Nevertheless, it continues to dominate popular imagination.

Alongside this veneration of fallen heroes was another, less heroic consequence of the war: one that aroused responses that varied from pity to embarrassed avoidance. It was the presence of thousands of young men, traumatised and made prematurely old by their experiences. Around two thousand returned soldiers spent the rest of their lives in repatriation hospitals. Nearly fifty thousand veterans were still

---

9   *The Age,* 26 April 1919, 7.
10 Charles Bean, *Anzac to Amiens* (Canberra: Australian War Memorial, 1983), 181.
11 Manning Clark, *A History of Australia,* Volume 6 (Melbourne: MUP, 1987), 120.
12 Clark, *A History of Australia,* 16.
13 Tom Frame, *Losing My Religion – Unbelief in Australia* (Sydney: UNSWP, 2009), 190.
14 Marilyn Lake and Henry Reynolds, *What's Wrong with ANZAC?: The Militarisation of Australian History* (Sydney: New South, 2010), vii.

in hospital in 1939; and in 1940, more than seventy thousand disabled men were receiving war service pensions.[15] Clark quotes Angela Thirkill who described the pathetic scenes of broken men, unable to express their inner trauma – except to each other – treated with embarrassed silence and often avoided by those who had not gone to war:[16]

> Richard Kirby, a pupil at the King's School, Parramatta, joined his classmates crowding round a war hero, an old boy of the school who had won his school colours in the Rugby XV in 1917. They put questions to the returned soldier. To Kirby's surprise the soldier replied in incoherent whispers. He had been gassed. All over the country young boys and girls saw ex-diggers break down and cry when they met each other, and only managed to say to each other the words of consolation: 'I know, Charl. I know'. Children came suddenly on grown-up men in distress and were told: 'He's ... crying. Can't stop ... Don't be frightened'.[17]

Les Carlyon suggests that the return of so many men who came home 'broken and bitter', and others who 'turned drunk and violent' may explain why many Australians 'in the thirty years after 1918 did not see the war, and Gallipoli in particular, in the romantic lights that have flickered around it in the new century'.[18] He describes the Great War as the 'worst trauma of the twentieth century for Australia ... a generation had lost many of its most generous male spirits'.[19] Many thousands of young women were to remain single and childless. It was indeed a catastrophe of unimaginable proportions.

Homecoming

It was to this war-weary nation that the twenty-seven chaplains of the Other Protestant Denominations, along with their brothers from the

---

15 Gammage, *The Broken Years*, 283.
16 Angela Thirkill, *Trooper to the Southern Cross* (Melbourne: Sun Books, 1966), 172-174.
17 Clark, *A History of Australia*, 117.
18 Les Carlyon, *The Great War* (Sydney: Macmillan, 2006), 753.
19 Carlyon, *The Great War*, 755.

major churches, returned to pick up their lives and ministries from where they had left them – some, like Frederick Miles and George Walden, nearly six years earlier. They returned to a nation that was emotionally and spiritually exhausted. Australia had suffered the highest casualty rate among the British Empire forces, and only a quarter of the original force of thirty thousand had survived.[20] They returned as heroes – part of the Anzac Legend – institutionalised in the creation of the Returned Services League,[21] 'which lost no chance in reminding people what a debt Australia owed to them'.[22] Even so, the cheers were beginning to die away before the last veterans reached home. The nation was weary of war and recoiling from its horror. People wanted to forget those tragic years.[23]

The chaplains played an important part in the establishment of war commemorations – especially the Anzac Day dawn service.[24] They also returned with an enormous admiration for the spirit of the men with whom they served. McKernan argues that many of them, having discovered how ignorant – and sometimes dismissive – the men were of Christian precepts, now recognised this spirit as a morality that 'existed independently of the churches'.[25] Others, especially evangelicals who reject the idea that people are Christians by virtue of their nationality, probably did not find this surprising but attributed such morality to the leavening effect of the gospel over centuries of European history. They would have agreed with Latourette's conclusion about the influence of Christ in history that 'No life ever lived on this planet has been so influential in the affairs of men...[and] has been the most fruitful source of movements to lessen the horrors of war'.[26]

---

20 Carlyon, *The Great War*, 752.
21 Originally named The Returned Sailors and Soldiers Imperial League of Australia.
22 Martin Crotty, 'The Anzac Citizen: Towards a History of the RSL,' *Australian Journal of Politics and History*, (2007), 190.
23 Gammage, *The Broken Years*, 270.
24 Michael Gladwin, *Captains of the Soul. A History of Australian Army Chaplains* (Sydney: Big Sky, 2013), 99.
25 Michael McKernan, *Australian Churches at War: Attitudes and Activities of the Major Churches 1914-1918* (Sydney: Catholic Theological Faculty and Australian War Memorial, 1980), 177.
26 K.S. Latourette, *History of the Expansion of Christianity*, Vol 7 (London: Eyre & Spottiswoode, 1945), 503-504.

The Church they returned to, however, was a sad reflection of the non-sectarian environment they had experienced at war. Serving alongside chaplains from other denominations amidst the misery and terror of the front line had given them a deeper appreciation of men they might previously have distrusted or even despised. McKernan refers to Presbyterian padre Alexander Stevenson, who wrote of a Roman Catholic chaplain, who had become his closest friend and had been killed in action: 'I shall never believe that a church which can produce such men is altogether evil'.[27] Similarly, the Baptist Frederick Miles described the Salvationist William McKenzie, whose presence on the first Anzac convoy he had initially considered 'absurd',[28] as 'a brother who has rendered magnificent service, a chaplain of sterling worth'.[29] Even before their return Alexander Main, editor of the *Australian Christian*, was pleading for a reconstruction in church life worthy of the returned soldiers in which the denominational lines would 'not be re-drawn but with-drawn'. He declared the failure to realise it would be 'criminal folly'.[30] But the bitterness of the conscription debates, which had led to a marked increase in sectarianism, guaranteed that Main's plea fell on deaf ecclesiastical ears. The trans-denominationalism the chaplains had experienced in the trenches struggled to survive with the coming of peace.[31]

One expression of this, which inflamed the latent feeling within OPD circles of being disregarded in favour of the major churches, is seen in a letter published in the *Australian Baptist* and re-published in the *Australian Christian* entitled 'A Glaring O.P.D. Injustice'. It argued that the Australian military authorities were discriminating against the Other Protestant Denominations in the matter of appointing chaplains to the ships bringing the troops home. Of one hundred chaplains to be assigned 'the Anglican Church was asked to supply 51; the Roman Catholics "all their available chaplains"; the Presbyterians, 18; the Methodists, 10; the O.P.D.s ONE! Are we content to be treated this way?' it thundered.[32]

---

27 *Presbyterian Messenger*, 20 October 1916; McKernan, *Australian Churches at War*, 139.
28 See p 33.
29 *Australian Baptist*, 16 May 1916, 2.
30 *Australian Christian*, 19 April 1917, 229-230.
31 McKernan, *Australian Churches at War*, 178.
32 *Australian Christian*, 6 March 1919, 138.

Miles responded to this with a long letter to the *Australian Christian* strongly disagreeing with the sentiments expressed. He stated quite emphatically that in the AIF the Other Protestant Denominations were 'dealt with in perfect equity' and received 'every courtesy and consideration'. With respect to the allocation of chaplains on home-bound transports he explained that each transport or hospital ship was to have one Roman Catholic and one Protestant chaplain, and the latter were to be appointed 'according to the percentage of the population of the Commonwealth claiming allegiance to the respective denominations'.[33] He then referred to the fact that a number of OPD ministers, who had been serving in the AIF as ordinary soldiers, had subsequently been commissioned as chaplains for troopship service, thereby negating the need for the Other Protestant Denominations to appoint more from within Australia.[34]

The editor of the *Australian Christian* concluded that Miles' explanation probably did give the correct interpretation to this alleged 'injustice' and that the Other Protestant Denominations had not been unfairly treated after all. He then side-stepped any accusation of unfair bias by adding: 'The *Australian Baptist*, the organ of the denomination to which Senior Chaplain Miles belongs, printed a paragraph which we republished – that is all'.[35]

Nevertheless, this incident shows that feelings of marginalisation were still very real within the OPD community. Despite the proud reputation their chaplains had established, there was still a perception of being seen as 'an aggregate of quite non-coherent and non-corporate factors' whose clergy represented something less than those of the major churches. A mere four years had passed since Archbishop Riley had stated publicly that: 'he could not sanction on behalf of the Church of England that a Baptist or Church of Christ minister should be officially in charge of Church of England men'. The memory of it, evidently, still rankled.

It is probably true, however, that by the end of the war the initial reluctance of the Chaplains General to accept the legitimacy of OPD

---

33 *Australian Christian*, 10 July 1919, 399.
34 *Australian Christian*, 10 July 1919, 399.
35 *Australian Christian*, 10 July 1919, 399.

chaplains as equal to theirs had changed. They appear to have accepted the Department of Defence's 1914 decision to adopt a policy of strict proportionality[36] – which was probably designed more to avoid potential sectarian squabbles by the churches than from ideological reasons on the part of Defence – as a fact of life and were prepared to live with it, as long as their share of the total chaplaincy establishment was maintained. The Other Protestant Denominations were now part of the *status quo* and would remain so providing they kept their place.

As for the OPD chaplains themselves, they came back older and very much wiser. The patriotic fervour and optimism of 1914 was now a distant memory. Having lived through the misery of the trenches and seen first-hand the horror of the casualty clearing stations, they no longer spoke glibly of the 'God of Battles' as they once may have. Porter notes how the war 'disturbed prophets and clerics who had welcomed war as an act of God, ordained by providence as salvation from a godless age'.[37] McKenzie probably spoke for them all when he described war as '"Hell" & no adequate description can picture its ghastliness'.[38] Michael Petras, similarly, compares the triumphalist Baptist preaching of 1914 with the disillusionment that followed:

> Baptist faith in the efficacy of sacrifice for the nation did not stand up to the reality of the horrors of modern warfare. Although the theme of sacrifice, as preached by the Rev. Peter Fleming and others early in the war, seems to have motivated many young men, it had lost most of its meaning by the time all was quiet on the Western Front.[39]

As early as 1917, evangelical chaplains had seen their own doubts expressed in the questions being put to them by those same young men who had originally responded to sermons about 'the God of Battles' and now asked, 'How can God let it go on? And how can God have willed

---

36 Gladwin, *Captains of the Soul*, 33.
37 Porter, 'The Sacred Service,' 23.
38 McKenzie, *Diary*, 69.
39 Michael Petras, 'Australian Baptists and The First World War in Retrospect,' in *Australian Baptists and World War I*, ed. Michael Petras (Sydney: Baptist Historical Society of N.S.W., 2009), 50.

all this?'[40] Michael McKernan argues that it was this earlier naïve and uncritical acceptance of the war as having been sent by God to 'chastise the people and alert them to the true path of devotion and duty' that, rather than raising the clergy's credibility, actually diminished it in the eyes of the population generally, especially by the war's end.[41] This seems to have been especially true for the Congregational Church, whose decline in membership increased after the war.[42] Hugh Jackson argues that the 'War, far from revivifying faith, sorely tried it, for, as the slaughter proceeded, many prayers went unanswered'.[43]

McKernan claims that 'many of the long-serving chaplains returned to Australia disturbed by the challenge of the AIF' no longer able to fit comfortably into the prevailing culture of their churches, and that some of them left the ministry.[44] This may have been true of Ashley Teece, who resigned from the pastorate on 20 April 1920 and was appointed one of the commissioners and deputy chairman of the original Repatriation Commission. He was later appointed Federal Commissioner for Repatriation.[45] The popular tabloid *Smith's Weekly*, which regularly advocated on behalf of returned servicemen, was unimpressed by Teece's appointment and on several occasions took him to task for being unsympathetic to veterans' claims for pensions. It reported that the congress of the Returned Services League in 1921 had invited him to 'face a resolution declaring that he (Teece) "no longer retains the confidence of returned soldiers."'[46] Four years later, in an article titled 'Teece, Perfect Teece', it referred to 'soldiers all over the Commonwealth whose pensions have come under the knife of this former servant of the church'.[47] Even more scathing was an article titled 'From Soul Saving to Sarcasm' which reported:

---

40 Robert Linder, *The Long Tragedy: Australian Evangelical Christians and the Great War, 1914–1918* (Adelaide: Open Book, 2000), 157.
41 McKernan, *Australian Churches at War*, 176.
42 Hugh Jackson, 'Aspects of Congregationalism in South-Eastern Australia, circa 1880–1930', Unpublished PhD dissertation, ANU, 1978, 91.
43 Jackson, *Aspects of Congregationalism in South-East Australia*, 146.
44 McKernan, *Australian Churches at War*, 177.
45 *The Age*, 27 August 1943, 3.
46 *Smith's Weekly*, 3 December 1921, 3.
47 *Smith's Weekly*, 20 December 1925, 11.

Before the war pushed Rev Ashley Teece into a lucrative job on the Repatriation Commission, he held down a pulpit in Adelaide. As a professional soul-saver, he preached Christianity for a pittance. When opportunity offered to drop the pulpit and practise cyaniding at a big salary, the bonds that bound him to the old life snapped.[48]

Smith's Weekly may have been unfairly biased against him, but its criticisms do raise the question whether Teece, who before the war had been quite a fervent evangelical, had been adversely affected by his experience and lost the passion that had previously energised his ministry.

The same does not appear to be true of most of the OPD chaplains, including fellow Congregationalists like Harold Perkins, who went to Samoa to serve with the London Missionary Society;[49] Frank Dowling, who directed the Society's Australian activities and was appointed President of the Congregational Union of Australia and New Zealand;[50] William Gunson, who became President of the Tasmanian Council of Churches and the Congregational Union of Tasmania;[51] Theodore Robertson, who served as Chairman of the Congregational Union of New South Wales;[52] and Reginald Turner, Superintendent of the Home Missions Board of the Congregational Union of New South Wales.[53] Two of them, Perkins and Robertson, continued their chaplaincy service in the CMF and were later appointed military district Senior Chaplains.

Of the other chaplains featured in chapter three, mention has already been made of Frederick Miles' awards for gallantry and service.[54] He was the longest serving OPD chaplain and served five years and nine months[55] - the only Senior Chaplain to serve throughout the entire war. Following the Armistice, he represented Australia at public functions in St Paul's Cathedral and Westminster Abbey and delivered the address at the first great Commonwealth Service in the Central Hall, Westminster.

---

48 *Smith's Weekly*, 9 February 1924, 27.
49 *West Australian,* 17 April 1935, 23.
50 *Brisbane Courier*, 26 April 1930, 23.
51 *Mercury*, 14 May 1942, 2.
52 *Sydney Morning Herald*, 26 October 1924, 8.
53 *Sunday Times*, 27 January 1929, 23.
54 Miles F.G Military Record, NAA B2455, 21, 36.
55 Miles F.G Military Record, NAA B2455, 39.

He also published a book titled *Triumph for the Troops*, for which Earl Birdwood wrote the foreword.[56] Miles returned to Australia in 1920 and spent three months travelling and lecturing on 'Glimpses of Gallipoli and Chaplaincy Cameos'.[57] He then returned to England and was appointed President of the Christian Endeavour Union of Great Britain and Ireland, and General Secretary of the Russian Missionary Society.[58]

William McKenzie arrived back in Australia on 10 January 1918[59] with his health shattered.[60] The 4th Battalion farewelled him at a special parade held in his honour, where the Brigade Commander spoke of his 'sublime and selfless service', after which, at the Battalion Commanding Officer's invitation, he passed up and down the assembled ranks shaking hands with every man.[61] Michael Gladwin records how more than seven thousand people, including many disfigured veterans, attended his welcome home concert at the Melbourne Exhibition Building, including one mother who travelled five hundred kilometres to 'kiss the hand of the man who had pieced together and buried her son's remains'.[62] The *Horsham Times* (he had previously ministered in Horsham), announcing his welcome home, reported:

> No man is better known among Australian soldiers at the front than Chaplain-Lieutenant Colonel W. McKenzie ... He has been with them in their trials and tribulations since the historic landing at Gallipoli: cheering them on their way, giving succour to wounded ... and even with his own hands laying to rest the bodies of men who have fallen on the battlefield.[63]

In 1926 he went to China to lead the Salvation Army's work, returning later to Australia, where he was promoted to the rank of Commissioner.[64] Fellow Salvationist, Benjamin Orames, having served for three years as

---

56 Miles F.G Military Record, NAA B2455, 102.
57 Miles F.G Military Record, NAA B2455, 104.
58 *The Age*, 8 July 1939, 26.
59 McKenzie W. Military Record, NAA B2455, 7.
60 *Sun* (Sydney), 24 April 1972, 13.
61 Woodbury, David. 'Do You Think I'm Afraid to Die with You,' *Halleluiah*, Volume 1, Issue 3, Autumn, 2008,' 16.
62 Gladwin, *Captains of the Soul*, 76.
63 *Horsham Times*, 12 February 1918, 5.
64 Woodbury, 'Do You Think I'm Afraid to Die with You,' 16-17.

Chief Secretary of the Salvation Army's Southern Territory, followed him as head of the work in China.[65]

George Walden continued to serve until his discharge on 10 May 1920.[66] Not only was he the oldest of the OPD chaplains but having served for five years and two months he was the second longest serving. He spent approximately thirty months at the Front (including one month at Gallipoli), which was longer than any fellow OPD chaplain, including McKenzie, who was at the Front for approximately twenty-eight months. This was a considerable achievement for a man of his age.

In 1919 Walden went with the occupying forces into Germany[67] and played a key role in the demobilisation of the Australian forces.[68] He was advanced to Chaplain Class 2 (Lieutenant Colonel) on 17 March 1919,[69] and Mentioned in Despatches on 11 June.[70] In November 1918 while still in France, he was invited by cable to become Secretary of the Churches of Christ Foreign Missions Board, an invitation he enthusiastically accepted.[71] On his homeward journey he visited the Churches of Christ mission work in India,[72] the first exercise of his new ministry which continued until poor health forced his retirement in 1934. The influence of his war-time experience carried through to his new role, and military imagery began to show in his writings, as can be seen in the following report to the *Australian Christian,* cited as having been received from Lt. Col. Chaplain G.T. Walden:

> Our missionaries in India and China are finding their forces enriched ... If we were to take from our Australian Churches' army and our Foreign Missionary forces the men and women trained at our Bible College, what a tragedy of loss we should experience ... we do not forget the army of workers making good in our Australian churches and on our foreign fields.[73]

---

65 *Register*, 28 April 1930, 5.
66 Walden G.T. Military Record, NAA B2455, 44.
67 *Australian Christian*, 14 August 1919, 471.
68 *Australian Christian*, 19 June 1919, 358.
69 Walden G.T. Military Record, NAA B2455, 9.
70 Walden G.T. Military Record, NAA B2455, 23.
71 *Australian Christian*, 15 April 1920, 154.
72 *Australian Christian*, 12 February 1920, 82.
73 *Australian Christian*, 16 September 1920, 418.

Though having been discharged from the AIF, he continued his military chaplaincy providing support at CMF camps. In 1929 he was reported as being on duty at the Gawler Infantry Camp from March 6 to 13.[74] He died on 26 July 1940,[75] a highly esteemed leader within Churches of Christ.

Arthur Forbes also returned to active ministry at the Auburn Church of Christ in Sydney. He continued his military chaplaincy in the CMF and was appointed Senior Chaplain for 2nd Military District. In September 1941 he returned to full-time military chaplaincy and served within Australia until 16 January 1944 when he retired from the Army and went back to civilian ministry at the Hamilton Church of Christ in Victoria. He was a veteran of three wars.[76]

George Cuttriss and Henry Procter both returned to pastoral ministry. Cuttriss, who was discharged on 12 February 1919, resumed his ministry at Hindmarsh in Adelaide, and Procter, who was discharged on 1 August 1920, returned to his ministry at North Richmond in Melbourne.[77] Contrary to McKernan's claim that many chaplains were 'disturbed by the challenge of the AIF', it seems, if anything, that it actually enhanced the later ministries of many OPD chaplains.

War Weariness

The communities to which they returned were weary of war and a spirit of revulsion swept the nation, brought about by the appalling casualty lists and the ever-present scenes of shattered bodies and minds. Governments saw few votes in defence issues, especially the Labor Party which, as Jeffrey Grey claims, 'professed a belief in collective security through the League of Nations but which opposed any measures which might create military forces capable of implementing it'.[78] The outcome was a widespread running-down of the machinery of national defence.

---

74 *Australian Christian*, 7 March 1929, 152.
75 *Australian Christian*, 31 July 1940, 466.
76 Arthur Bottrell. 'Forbes, Arthur Edward (1881-1946),' *ADB*, Volume 8 (Melbourne: MUP, 1981), 539.
77 Nutt, Dennis. 'Military Chaplains: For Service of our Soldiers', 27.
78 Jeffrey Grey, *A Military History of Australia* (Melbourne: CUP, 2008), 123.

The AIF was replaced on 1 April 1921 by a reorganised militia whose total war establishment was set at two hundred and seventy thousand, but whose strength in 1921 only stood at one hundred and twenty-seven thousand, declining to thirty-seven thousand in 1922.[79] Similarly, the Navy was significantly reduced in size. Immediately after the war it had five thousand three hundred and fifty personnel manning one battlecruiser, three light cruisers, twelve destroyers, four sloops, six submarines and a number of auxiliary vessels. But by 1933 the number of personnel had fallen to just under three thousand men, 'whose main purpose was to maintain a cruiser force large enough ... to maintain coastal defence'.[80]

The one new development was the formation of the Royal Australian Air Force, which grew out of the AIF's Australian Flying Corps. During World War I, eight hundred officers and two thousand eight hundred and forty men served in the AFC and one hundred and seventy-five lost their lives. Chief among the AFC veterans who helped to lay the groundwork for the new Royal Australian Air Force was Sir Richard Williams, 'affectionately regarded as the "Father of the Air Force", who became Chief of the Air Staff'.[81] In January 1920 the AFC was replaced by the Australian Air Corps, which became the Australian Air Force on 31 March 1921. Then, having received the King's consent on 13 August 1921, it became the Royal Australian Air Force. It had a strength of twenty-one officers and one hundred and thirty other ranks and was equipped with one hundred and sixty-four aircraft, one hundred and twenty-eight of them gifted by the British government.[82]

Army Chaplaincy Between the Wars

The huge reduction in defence spending, which accelerated during the Great Depression, seriously affected the functioning of chaplaincy in

---

79 Grey, *A Military History of Australia*, 125.
80 Rowan Strong, *Chaplains in the Royal Australian Navy, 1912 to the Vietnam War* (Sydney: UNSW Press, 2012), 125-126.
81 Peter A. Davidson, *Sky Pilot, a History of Chaplaincy in the RAAF 1926–1990* (Canberra: Directorate of Departmental Publications, Department of Defence), 1-2.
82 Grey, *A Military History of Australia*, 132; Air Force, Our Journey, https//www.airforce.gov.au/about-us/history/our-journey.

both Army and Navy. Douglas Abbott notes that the Army Chaplains Department virtually ceased to exist between 1919 and 1939. The Chaplains General had very little scope for action apart from providing a link between the Army and the governing bodies of their churches. Apparently, they were little more than figureheads, sometimes not being consulted in the appointment of new Senior Chaplains. There was no dedicated chaplains' staff system at Army Headquarters, and the only chaplaincy structure that remained functioned at Military District level where Senior Chaplains arranged the supply of chaplains for annual camps at the request of the local Military District Headquarters.[83] Tom Johnstone enlarges on this, reporting that 'for long periods between the two world wars there were no full-time army chaplains', and that Army Headquarters in 1933 had to inform Military District commanders and Senior Chaplains that 'facilities for religious practice' were not being made available at some training camps.[84]

The Army itself must bear some of the blame. There appears to have been a poor appreciation of the value of chaplaincy in the peacetime CMF. An undated typescript entitled 'AAChD Between Wars',[85] held by the Australian War Memorial, indicates that chaplains were the only people attending annual camps who were unpaid, merely drawing field allowance. They were not issued with basic items like haversacks and water bottles, and rarely was tentage made available for them. Neither were those appointed to cavalry or artillery units provided with horses. Furthermore, the Army refused to post chaplains officially on the strength of CMF units – except for duty at annual camps – making it impossible for chaplains to form the sort of relationships with units that had proved to be so important during World War I. Any extra services that a chaplain might provide to a unit had to be done unofficially and without any form of payment.[86]

Of particular interest is the apparent breakdown of the policy of proportional representation of chaplains according to denominational size.

---

83 Douglas Abbott, 'In This Sign Conquer: The Chaplains General of the Australian Army, 1913–1981'. Unpublished manuscript, 1995, 59.
84 Tom Johnstone, *The Cross of Anzac, Australian Catholic Service Chaplains* (Brisbane: Church Archivists' Press, 2003), 89.
85 Australian Army Chaplains Department.
86 'AAChD Between Wars', undated MS typescript, AWM 54, 177/2/1 1.

The aforementioned typescript described the authorised chaplaincy establishment in each state as 'nonsensical,' reporting that the Salvation Army, whose national membership comprised less than one percent of the population, actually had as many chaplains as the Anglicans who, according to the National Census, represented forty-six percent.[87] This again raises the question of the relevance of the system of proportionate representation in an environment that for the most part requires a trans-denominational chaplaincy service. It is especially so when particular denominations are unable to provide the number of chaplains they claim as their entitlement, as appeared to be the case here.

The 'rudimentary structure' provided by the Senior Chaplains in each military district, which managed the day-to-day functioning of chaplaincy, was eventually upgraded with Army Headquarters appointing Senior Chaplains to act and sign as deputies for their respective Chaplains General. This development was an important step in the Other Protestant Denominations' journey toward equal standing with the major churches in that an OPD Senior Chaplain was included, even though there was as yet no OPD Chaplain General. It was a further indication of its emerging recognition as an integral part of the chaplaincy structure. This appointment had oversight of chaplains from the Baptist, Congregational and Churches of Christ denominations, but not the Salvation Army which, though part of the group during World War I, after the war functioned separately until 1942.[88]

The United Board

February 1920 saw a significant change when the designation *Other Protestant Denominations* was changed to United Board (UB) to bring it into line with its British counterpart. The *Daily Telegraph* reported:

> The Defence Department has ... officially notified an alteration relative to the present designation of O.P.D., "Other Protestant Denominations," of the chaplains' department, Australian

---

87 'History of Army Chaplains,' AWM 54, 177/2/1/1 1.
88 Gladwin. *Captains of the Soul*, 90.

Military Forces. The present senior O.P.D. chaplains will henceforth be designated "Senior Chaplains, United Board," and Commandants of military districts have been advised accordingly.[89]

The May edition of the *Australian Christian* also noted this change:

Changes of Designation. – The Senior Chaplains 'Other Protestant Denominations,' will in future be designated 'Senior Chaplains, United Board'.[90]

The United Board in each State continued to do what its predecessor had done since 1913: relating to the relevant Military District Commandants in nominating both United Board Senior Chaplains and new chaplains, maintaining pastoral oversight of its chaplains and reinforcing the important concept that chaplains are primarily ministers of the Church 'on loan' to the armed services.

A number of UB chaplains who had served with the AIF continued serving in the post-war CMF, including Forbes, Perkins, Robertson and Walden, all of whom became Senior Chaplains. However, the New South Wales Baptist Assembly in 1919 'adopted a recommendation from its current chaplains that the existing chaplaincy appointments be terminated and that returned soldiers [who were already ordained ministers] be appointed in their place'.[91] While this deprived the chaplaincy team of much valuable experience, it injected 'new blood' and countered a later description of many chaplains being 'too old or too weak to undertake war service in 1939 [and] who should have retired before to allow younger men to be trained'.[92]

Another issue over which the Baptists disagreed with their United Board partners related to rank insignia. Gladwin notes that many chaplains during World War I identified the need for a clearer definition of chaplains' rank, which AIF regulations defined as being temporary

---

89 *Daily Telegraph*, 8 May 1920, 8.
90 *Australian Christian*, 20 May 1920, 184.
91 Bruce Thornton, Thornton, Bruce. *And it Brought Forth Fruit – A History of the Association of Baptist Churches NSW & ACT.* Boston: Greenwood, 2020, 75.
92 'AAChD Between Wars,' AWM 54, 177/2/1/1 1.

commissions with relative Army ranks and insignia.[93] The outcome was quite different from what most chaplains expected, and in May 1920 the *Australian Christian* reported:

> A recent military order reads as follows: – CHAPLAINS' DEPARTMENT: *Relinquishment of Military Rank*. – It is notified for information that all Chaplains of the Australian Military Forces will, from 30th April 1920, relinquish the military rank, whether substantive, honorary, relative, or temporary, hitherto held by them. They will, however, continue to be classified as Chaplains, 1st, 2nd, 3rd, or 4th Class, according to their length of service.[94]

Miles suggested this in late 1919 while still the AIF's OPD Senior Chaplain. The Adjutant-General sent it to the heads of the member denominations who, with the exception of the Congregational Union, the Churches of Christ and the Salvation Army, were in favour, as were the Chaplains General. Consequently, in March 1920 the Adjutant-General recommended the abolition of rank for Army chaplains. Despite the mixed feelings of the chaplains themselves, the majority of whom believed rank should be retained,[95] the Statutory Rules affecting chaplains (SR120/1920) were amended to say that chaplains would be officers but would not hold any rank other than Chaplain. The Defence Act was eventually amended accordingly in 1927.[96]

For the next two decades the United Board continued to represent its member churches as part of the 'haphazard' administration of Army chaplaincy. Its chaplains attended the biennial military training camps and performed the occasional duties required of CMF chaplains. It was the outbreak of World War II and the threat of a Japanese invasion that once again galvanised the Board into action and demonstrated what it could do to support the nation in the approaching crisis.

---

93 Gladwin, *Captains of the Soul*, 91.
94 *Australian Christian*, 20 May 1920, 184.
95 Gladwin, *Captains of the Soul*, 92.
96 Abbott, *The Chaplains General of the Australian Army*, 62; Johnstone, *The Cross of Anzac*; 89; The Defence Act 1927, para 1041.

Naval Chaplaincy Between the Wars

The Royal Australian Navy, though benefiting from a defence policy that prioritised naval expenditure, was also subject to stringent cost-saving.[97] Unlike the Army, the naval chaplaincy establishment remained much as it had been during the war. From 1924 chaplains were appointed initially for a two-year probationary period which, if successful, led to a confirmed appointment to a permanent position with a compulsory retiring age of forty-five.[98] The main change was that in 1921 a second full-time Roman Catholic appointment was approved in addition to the Roman Catholic Fleet chaplain.[99] This continued until 1929 when the Roman Catholic chaplain at Flinders Naval Base resigned. Added financial constraints during the Depression years meant that he was not replaced until 1938 when the establishment of five Protestant chaplains and one Roman Catholic was increased to six Protestants and two Roman Catholics.[100]

The first serious attempt by the United Board to be included in naval chaplaincy was in 1933 when Dr Ernest Watson, a prominent Baptist minister, on behalf of the United Board in New South Wales, offered his services as 'honorary chaplain to the Navy'.[101] The Captain Superintendent of Naval Establishments, Sydney, warmly supported Watson's application in a signal to the Secretary of the Naval Board mentioning that 'the United Board have official recognition in the Royal Navy'.[102] The Secretary replied expressing appreciation but said the Naval Board members 'regretted they were unable to avail themselves of Dr Watson's offer'.[103]

One year later another attempt was made by the Baptist Union of New South Wales, which wrote to the Naval Board informing it of a

---

97 Mike Carlton, *Flagship* (Sydney: Random House, 2016), 34.
98 Strong, *Chaplains in the Royal Australian Navy*, 128
99 Johnstone, *The Cross of Anzac*, 31.
100 Strong, *Chaplains in the Royal Australian Navy*, 147.
101 Chaplain RANR – Appointment of One Representing United Board, AWM MP150/1, 431/202/187, 12.
102 Chaplain RANR – Appointment of One Representing United Board, AWM MP150/1, 431/202/187, 11.
103 Chaplain RANR – Appointment of One Representing United Board, AWM MP150/1, 431/202/187, 8.

resolution that was carried at its Annual Assembly on 26 September 1934 which read:

> That this Assembly views with concern the fact that the United Board has no chaplain in the Australian Navy and Air Force, and urges the Federal authorities to make some arrangement whereby an appointment may be made.[104]

The Naval Board's reply explained that the number of Protestant chaplains required was small and the current establishment of chaplains was filled by chaplains from the Church of England, Presbyterian and Methodist denominations, who minister to all Protestants on ships or shore establishments, and that 'the number of serving members belonging to the denominations represented by the United Board would not at present justify the appointment of a United Board chaplain'.[105] It was clear that the Navy still maintained an uncritical acceptance of the system of proportionate representation even though it had initially wanted a trans-denominational system. A further attempt was made in April 1937, to fill a vacancy left by the resignation of a Methodist reservist chaplain at Williamstown. Once again it was unsuccessful. The 2nd Naval Member of the Naval Board[106] declared: 'The present members and I are not prepared to amend the existing practice of nominations only by the Protestant Chaplains Nominating Committee'.[107] Consequently, the denominational *status quo* that had been in place since the birth of the Royal Australian Navy in 1912 remained throughout World War II and continued until the late 1960s.

Chaplaincy in the Royal Australian Air Force

The first official reference to chaplains in the Air Force is dated 19 January 1926, five years after its birth, when Number 1 Squadron based

---

104 Chaplain RANR – Appointment of One Representing United Board, AWM MP150/1, 431/202/187, 7.
105 Chaplain RANR – Appointment of One Representing United Board, AWM MP150/1, 431/202/187, 5.
106 The Naval Board governed the RAN from its inception until the end of World War II.
107 Chaplain RANR – Appointment of One Representing United Board, AWM MP150/1, 431/202/187, 1.

at Laverton, Victoria, sought advice from RAAF Headquarters about the tenure of chaplains. At that time the RAAF had no established procedure for appointing chaplains although, according to Peter Davidson, some base commanders had made their own *ad hoc* appointments of honorary civilian clergy.[108]

Then, on 14 June the Air Board presented a submission to the Government arguing for a chaplaincy service, whose members would be part of the Citizen Air Force and whose service would not exceed twenty-five days per year. It was agreed that RAAF chaplaincy would adopt the original Army rank system of four classes of chaplain whose worn rank would be flight lieutenant through to group captain (even though the Army at that time had temporarily abandoned the practice).[109] The outcome was the appointment on 24 March 1927 of two Anglican and two Roman Catholic nominees. Rev Frederick Hughes and Fr Walter Walsh at Point Cook, and Rev Oswald Dent and Fr Richard Darby at Richmond became the RAAF's first commissioned chaplains, their commissions being made retrospective to 1 March 1927.[110]

For the first twelve years all RAAF chaplains served part-time, but by 1936 Williams was considering the need for a full-time chaplain. However, it was not until 4 May 1939 that George McWilliams (Anglican) was appointed the first full-time chaplain on the Permanent Air Force list. The decision to appoint an Anglican was based on the fact that forty percent of serving members of the RAAF were listed as Anglicans and, under the system of proportionate representation, it seemed appropriate.[111] As with Navy and Army chaplaincy it set a precedent that continued for decades. Davidson reports that from 1926 until the outbreak of World War II the only denominational representation was eleven Anglicans, eight Roman Catholics and four Presbyterians.[112]

The first approach to the RAAF by the United Board followed

---

108 Davidson, *Sky Pilot*, 1-2.
109 Air Board Agenda 767 (RAAF) – Conditions of Service in the Chaplains Branch, NAA A14487, 7/AB/767, 1-4.
110 Davidson, *Sky Pilot*, 1-3; Air Board Agenda 767 (RAAF) – Conditions of Service in the Chaplains Branch, NAA A14487, 7/AB/869, 1-2.
111 Davidson, *Sky Pilot*, 1-8; Air Member for Personnel – Appointment of a Permanent Chaplaincy Policy, NAA A705, 36/1/98; A705, 36/1/111.
112 Davidson, *Sky Pilot*, 1-10; Appointment of Anglican Replacements, NAA A14487, 9/AB/1500; A14487, 9/AB/1478.

the previously mentioned resolution of the New South Wales Baptist Union Annual Assembly expressing its concern that there was no United Board chaplain in either Navy or Air Force.[113] Again the response was that there were no vacancies and insufficient adherents of the United Board denominations to warrant a chaplain. It was the outbreak of war and the rapid expansion of the RAAF that, as with the 1st AIF, exposed the lack of logic in a policy that tied what was essentially a trans-denominational ministry to a denominational selection process, and finally opened the door to United Board participation.

The Gathering Storm

Reflecting on the years between the two world wars, Sir Winston Churchill named the first volume of his history of World War II, 'The Gathering Storm'.[114] It was an appropriate title for a book whose purpose was to describe how 'the English-speaking peoples, through their unwisdom, carelessness and good nature allowed the wicked to rearm'.[115] Australia was one of those nations. Its military chaplaincy during the twenty years between the two world wars, like the armed services it served, was minimal and ill-prepared for the catastrophe that lay ahead.

The veterans of the 1st AIF, including its twenty-seven OPD chaplains, returned home to pick up their lives and ministries – many of them going on to make significant contributions to the post-war world and church. The nation, though intensely proud of them and what they had accomplished was sick of war. Indeed, the memory of it haunted people's minds well into the 1930s. Peter Firkins reflects on their 'half-despairing hope', as the sufferings of the Great Depression were beginning to fade, that they would not have to suffer it again 'and would be allowed to enjoy a modest contentment … until, almost too late, they realised that they must [once again] prepare for war'.[116]

---

113 Chaplain RANR – Appointment of One Representing United Board, AWM, MP150/1, 431/202/187, 7.
114 Winston S. Churchill, 'The Gathering Storm,' Vol 1, *The Second World War* (London: Casssell, 1948).
115 Churchill, 'The Gathering Storm,' ix.
116 Peter Firkins, *The Australians in Nine Wars, From Waikato to Long Tan* (Sydney: Pan, 1982), 185.

The United Board had made a few unsuccessful attempts to gain access to the chaplaincy branches of the Navy and the Air Force. But the reality was that, despite its hard-won acceptance into the wider world of Australian military chaplaincy, it remained a relatively insignificant provider of chaplains, whose contribution basically was occasional chaplaincy support at CMF camps. A war-weary nation in a war-weary world had little enthusiasm for building on or even adequately maintaining what it had possessed in 1919. And so, the once rejected stone was largely put into storage, ready for another crisis that everyone hoped – and initially believed – would never come.

But it did. A.K. MacDougal, commenting on the rise of Fascism in Italy, Hitler in Germany and Japanese atrocities in Manchuria, described the 1930s as 'a decade of international gangsterism,' which eventually in 1938 caused the Australian government to announce that 'an unprecedented forty-three million pounds would be spent on Defence over the next three years'. The government also called for volunteers for the CMF, or the Militia as it was commonly known, causing it to expand to around eighty thousand by 1939, in addition to the three thousand trained troops in the permanent army.[117]

Then, on 3 September 1939, Prime Minister Robert Menzies, in a radio broadcast to the nation, informed his fellow Australians that it was his:

> Melancholy duty to inform [them] officially that in consequence of a persistence by Germany in her invasion of Poland, Great Britain has declared war upon her and that, as a result, Australia is also at war.[118]

The nation's worst fears had been fulfilled and the United Board was about to be brought out of storage to face its biggest challenge.

---

117 A.K. MacDougal, *Anzacs Australians at War* (Sydney: Currawong, 1994), 134.
118 *The Age*, 4 September 1939, 11.

CHAPTER FIVE

World War II: The Reshaped Stone

ON 1 SEPTEMBER 1939 A German invasion force crossed the Polish border, ending the fragile hope of 'peace for our time'[1] and launching the greatest conflict in world history. It was, as Max Hastings reflects, a war in which: 'Men and women from scores of nations struggled to find words to describe what happened to them'.[2] Two days later, Britain, following its pledge of support for Poland given on 31 March,[3] declared war on Germany. This announcement was followed thirty minutes later by a similar declaration from the Australian Prime Minister.[4] The *Sydney Morning Herald* reported:

> The British Prime Minister, Mr. Chamberlain, in a broadcast from No. 10 Downing Street at 8.15 p.m. Sydney time yesterday, said that Great Britain was at war with Germany ... At 8.45 o'clock last night, the Prime Minister, Mr. Menzies,

---

1 British Prime Minister Neville Chamberlain's famous declaration in London on 30 September 1938 following his Munich meeting with Adolf Hitler.
2 Max Hastings, *All Hell Let Loose, The World at War 1939-1945* (London: William Collins, 2012), xv.
3 Winston S. Churchill, 'The Gathering Storm,' Vol. 1, *The Second World War* (London: Cassell, 1948), 270.
4 Victor Swain, *Australia: Moments in History* (Sydney: New Holland, 2011), 281.

who is in Melbourne, announced that the Commonwealth of Australia was [also] at war with Germany.[5]

The German onslaught, using blitzkrieg tactics for the first time, was devastatingly effective. 'By the end of the second week of the campaign, the Polish army had ceased to exist as an organized force ... the guarantee of aid from the Western Allies had proved virtually meaningless'.[6] The war then, apart from the 'winter war' in Finland,[7] entered a brief period of calm, dubbed by the American press the 'phoney war', until the opening of Hitler's Western offensive in the following spring.[8]

It did, however, provide a temporary breathing space for Australia to address the serious shortcomings of its inter-war defence policies, making it 'possible for the 2nd AIF to be built up without any great feeling of urgency'.[9] It also galvanised Australia's 'haphazard' military chaplaincy systems. For the United Board it opened the way for what eventually became an unprecedented expansion following the entry of Japan into the war.

Four events that occurred between 1940 and 1942 largely made this possible, effectively re-shaping the United Board and bringing it into line with the major churches. These were: the appointment of a fifth Chaplain General; the appointment of a fifth Staff Chaplain for the Air Force; the addition of Lutheran chaplains to the group; and its evolution from State based boards into the Federal United Churches Chaplaincy Board. However, before considering these in greater detail it is important to understand the nature of the environment – ecclesiastical and national – in which they transpired.

---

5 *Sydney Morning Herald*, 4 September 1939, 1.
6 Ronald Heiferman, 'World War II,' in *Wars of the 20th Century*, ed. S.L. Mayer (Secaucus, N.J.: Derbibooks, 1975), 257-258.
7 A war between the Soviet Union and Finland that began on 30 November 1939 and ended with the Moscow Peace Treaty three and a half months later.
8 B.H. Liddell Hart, *History of the Second World War* (London: Pan, 1973), 36.
9 Peter Firkins, *The Australians in Nine Wars, From Waikato to Long Tan* (Sydney: Pan, 1982), 186.

The Churches' Response to the Declaration of War

Australians faced the prospect of a new world war with a sense of resignation quite different from the patriotic enthusiasm at the outbreak of World War I. Bitter experience had taught the nation that modern global wars were likely to be long and bloody. Nevertheless, there was widespread support for the government's 'melancholy duty', including from the churches, whose leaders reluctantly accepted that this war was a necessary evil, and to avoid it would bring upon the world a greater evil. The General Assembly of the Presbyterian Church 'unanimously agreed to a motion assuring the Commonwealth Government of whole-hearted support in the War'.[10] Likewise, Anglican Archbishop of Brisbane, John Wand, expressed the general feeling of church leaders when, in a radio broadcast on 28 December 1939, he described the war as a crusade:

> To defend the sacred places of the human spirit – the citadels of ideals and principles which have become sacred to us in the course of many centuries of Christian teaching and practice – against those who would eradicate them.[11]

It was a theme he continued to propound throughout the war, arguing that 'we are fighting for our laws and institutions ... [and] if we were beaten in this war, we should lose all these things'.[12]

As in World War I, there were pacifists who opposed Australia's involvement in the war, but they were a very small minority. Church leaders were overwhelmingly united in their belief that a war against Nazi Germany, whose belligerence, ideology and violation of human rights were already manifest, was not only justified but essential for the greater good of humankind. Archdeacon T.C. Hammond, Principal of the Anglican Moore Theological College, reflected the general opinion when he asked: 'Are we to contend that the unrestrained ambition and cold-blooded tyranny that brings subject people into intolerable bondage awakens no indignation in the Most High'?[13]

The outbreak of World War II, coming as it did so soon after the

---

10 *Sydney Morning Herald*, 17 May 1940, 3.
11 *Australian Christian Commonwealth*, 26 January 1940, 7.
12 *Courier-Mail*, 24 March 1941, 4.
13 *Church Record*, 12 February 1941, 2.

horrifying slaughter of World War I and the subsequent hope that it had been 'the war to end all wars', confronted church leaders with a major dilemma. Memories of 1914 and the churches' role as major recruiting agents for the Army and Navy were still fresh – sometimes embarrassingly so – in church leaders' minds. Nevertheless, it was clear to them that in the current situation they had little choice other than to support the Government. Ian Breward notes that the churches 'treated the declaration of war with far more sobriety than they had in 1914',[14] and Michael Gladwin concludes:

> Although church leaders were firm in their endorsement of the war, the horrendous human and material losses of the Great War meant that their patriotism was muted, and they were certainly not the 'eager recruiting agents' that the 1914–18 generation of Anglicans and Protestants had been.[15]

The member churches of the United Board were equally resolved to stand against the evil forces they saw at work in the world. In his first address to the Congregational Union following the declaration of war, John Rupert Firth, Chairman of the Union, regretting the 'chain of events which had led inevitably to the present war' asserted that 'We are fighting for all those things for which the Church stands – for the liberty of the spirit, for tolerance, honesty of dealing, and good faith'. He then warned against those who were not prepared to fight:

> Within our denomination are included people of almost every shade of political opinion, people with conservative ideas on social problems, people with radical ideas on social problems, and people with no ideas at all! You do not need me to tell you that the really dangerous section in a denomination like ours, and in a democracy like ours, is the apathetic section, who are prepared to see the liberty steadily whittled away, under their indifferent noses.[16]

14　Ian Breward, A *History of the Australian Churches* (Sydney: Allen & Unwin, 1993), 130.
15　Michael Gladwin, *Captains of the Soul, A History of Australian Army Chaplains* (Sydney: Big Sky, 2013), 101.
16　John Rupert Firth, 'Chairman's Address,' in *Year Book for 1940* (Sydney: Congregational Union of New South Wales, 1940), 31-33.

In similar vein, the *Australian Baptist* published an article by Rev Harold Dart expressing the Baptist attitude towards the war. Dart argued that contrary to the idealistic hopes that the creation of the League of Nations would bring an end to wars, the Bible taught that 'wars and rumours of wars' would continue until the return of Christ and that:

> Sin necessitates magistrates, police, army, navy and air-force. As long as sin remains, there will of necessity remain the sword of justice, to execute judgement against civil wrong. The sword of order to put down rebellion and civil commotion, and the sword of war to restrain the violence of external foes ... Obedience to Caesar is, within limits, a part of the will of God, and a direct command of Jesus Christ.[17]

The Churches of Christ, whose Victorian and New South Wales Conferences in 1931 had both adopted strong support for international efforts towards disarmament and the pursuit of peace through 'religious and ethical standards',[18] by 1939 had also accepted the bitter necessity of war. Alexander Main, in an article titled 'War Has Come Again,' called upon the church to exercise a ministry of comfort:

> A special call comes to the church. A day of testing lies ahead, and yet a day of opportunity. God's children can call upon men to find in God the one source of peace and security. The church has a ministry of comfort to perform. There are the men who will be drafted for service. They need Christ, and they should have all the moral and spiritual reinforcement we can give them. There are mothers and fathers and wives who will suffer loss, and great hosts of anxious minds and with hearts filled with sorrow. There are still the conditions which breed war to be removed and the forces of evil underlying it to be combated.[19]

---

17 *Australian Baptist*, 12 September 1939, 1.
18 *Australian Christian*, 9 April 1931, 213; 25 June 1931, 388.
19 *Australian Christian*, 6 September 1939, 562.

Mobilisation of the Nation

The first three years of the war saw a gradual mobilisation of the nation which began with an announcement on 15 September that a new division of troops would be raised for overseas service. Two days later it declared that a significant increase in training for the Militia was to take effect.[20] *The Age* reported:

> The task with which the Government is most immediately confronted is how to promote the efficiency of the militia forces ... The 80,000 militia are now to be called up in two successive batches in order to be trained in camps for two months.[21]

On 20 October the government went even further and reintroduced compulsory military training, requiring unmarried men turning twenty-one to undertake three months' training with the Militia.[22] Without it, Australia's citizen forces and its tiny regular army, lacking anti-aircraft guns, modern field artillery and armoured vehicles, was incapable of fighting a battle-hardened enemy.[23]

These decisions indicated that, as in World War I, Australia intended to continue the policy of maintaining two armies: one volunteer and eligible for overseas service, and the other for home defence. This had implications for the United Board which, as the previous chapter noted, had a significant number of clergy – some of them World War I veterans – serving as CMF chaplains. It reflected the strongly held Australian commitment to the concept of the citizen soldier, whose origin went back to Australia's first military leader, Sir Edward Hutton, and his advocacy of 'a national army of citizen soldiers'.

Nevertheless, it was an awkward system that differed from the British system of a Regular Army and a part-time Territorial Army. The first and second AIFs were made up of civilian soldiers who served under different obligations to those of the CMF. Jeffrey Grey argues that it was 'a short-sighted policy which caused great difficulty after

---

20 Michael McKernan, *The Strength of a Nation* (Sydney: Allen & Unwin, 2008), 31; Mark Johnston, *Anzacs in the Middle East* (Melbourne: Cambridge University Press, 2013), 1.
21 *The Age*, 18 September 1939, 8.
22 Compulsory Military Service, AWM 54, 946/2/1.
23 John Edwards, *John Curtin's War*, Volume 1 (Sydney: Viking, 2017), 173.

the beginning of the war in the Pacific...[causing] rivalry between the officers of the Staff Corps and those of the citizen forces'.[24] Anthony MacDougall agrees, commenting that the voluntary system which gave Australia 'fighting men like no other ... was essentially divisive'.[25] Later in the war these differences gradually disappeared as Militia and AIF soldiers came under one command and fought together in Papua New Guinea.

The United Board Prepares for War

The United Board was at one with the major churches in their commitment to provide adequate chaplaincy support for the rapidly expanding armed services. The period from the declaration of war to the rapid acceleration of mobilization in 1942, following Japan's entry into the conflict, became the period in which the re-shaping of the United Board took place, enabling it finally to take its place as an equal – though still junior – partner to those churches.

Initially they all easily met their quotas of chaplains based, as in World War I, on census figures. The build-up of military forces was gradual and did not reach its peak until August 1943.[26] Likewise, the build-up of chaplains from 1939 to the end of 1941 was also gradual. The immediate priority was the proliferation of training facilities, both Army and Air Force, within Australia. But it soon included chaplaincy to the 2nd AIF in the Middle East and airmen serving overseas. The Navy, numerically the smallest of the three services, received less attention.[27] Johnstone describes it as being 'virtually neglected',[28] although it did increase from nine to thirty chaplains over the course of the war.[29]

---

24 Jeffrey Grey, *A Military History of Australia* (Melbourne: CUP, 2008), 146-147.
25 A.K. MacDougall, *Anzacs Australians at War* (Sydney: Currawong, 1994), 139.
26 Grey, *A Military History of Australia*, 153.
27 Carlton, *Flagship*, 117f.
28 Tom Johnstone, *The Cross of Anzac, Australian Catholic Service Chaplains* (Brisbane: Church Archivists' Press, 2003), 91.
29 Rowan Strong, *Chaplains in the Royal Australian Navy, 1912 to the Vietnam War* (Sydney: UNSW Press, 2012), 195

Army Chaplaincy

The Army Chaplains Department quickly responded to the emerging international crisis as the Chaplains General informed the Senior Chaplains in each military district of the new chaplaincy establishment numbers within their jurisdictions. The immediate need was to recruit more chaplains to serve the Militia camps that developed around the major cities, and then to provide for the 2nd AIF as new divisions were raised.[30] Building on its corporate experience of war and following the established principle of appointment according to census figures, the Chaplains Department swung into action.

The Anglican allocation was questioned by Ernest Burgmann, Bishop of Goulburn, who claimed that almost three-quarters of those who enlisted in World War I were Anglicans, even though only half of the chaplains were. He recommended that the number of chaplains appointed by each denominational group should be in accordance with the size of the declared religious affiliations of men enlisting, and that consequently there should be an increase in the number of Anglican chaplaincy positions.[31] Defence, however, decided to maintain the principle of proportionate representation based on census figures.[32]

Nevertheless, the final figures, as recorded by Douglas Abbott,[33] reveal that of the nine hundred and seventy-two chaplains who served in World War II, one hundred and seventy-three were from the United Board.[34] This is more than three times the number allocated to them. The equivalent Anglican figure was three hundred and forty-one, Roman Catholic two hundred and six, Presbyterian one hundred and twenty-seven, and Methodist one hundred and twenty-one.[35]

---

30 Gladwin. *Captains of the Soul*, 102.
31 Burgmann, E., Bishop of Goulburn. Correspondence with Archbishop Wand, 6 October 1939, Anglican Archives, Brisbane.
32 Appointment of Chaplains, NAA MP508/1, 56/750/340.
33 These were based on denominational reports presented at the August 1946 Chaplain Generals' Conference and from Australian Army Chaplains Department records (1946).
34 Chaplain General Allen Brooke reported the total number to be 174, which included a Christian Science chaplain who, for administrative purposes, was included with the United Board; Allen Brooke, Letter to FUCCB, 23 July 1947.
35 Douglas Abbott, *In This Sign Conquer, The Chaplains General of the Australian Army 1913–1981*, Unpublished Manuscript, 85-87.

Furthermore, one hundred and four of the seven hundred and fifty-four[36] chaplains who served full-time were from the United Board.[37] In all there were fifty-two Baptists, forty-two Congregationalists, thirty-four Churches of Christ, forty Salvation Army, and five Lutherans.[38]

These statistics are quite illuminating in that they reveal the potential weaknesses of a proportionate system based on census figures. Though accurate in terms of nominal membership, census figures do not necessarily reflect the active strength of participating denominations. Nor do they assess their ability to meet the target goals. This was particularly true of the Anglicans whose total number of chaplains, though larger than any of the others, fell significantly short of the number that the census required. By the latter part of the war there simply weren't enough Anglican clergy young enough and fit enough to meet a requirement based on a figure that included a huge number of servicemen and women who were Anglican in name only. This disparity would have been even more serious if Defence had accepted Burgmann's recommendation that Anglican positions be increased. Roman Catholic, Presbyterian and Methodist figures equated more closely to census expectations, but the eventual size of the United Board contribution to Army chaplaincy is remarkable in relation to what officialdom originally envisaged, and later declared 'represents a worthy ministerial manpower contribution'.[39]

Air Force and Navy Chaplaincy

Air Force chaplaincy also began to expand significantly. When war was declared the Air Force only had six chaplains:[40] George McWilliams

---

36 Gladwin's research discovered two more chaplains than the 752 listed in 'Royal Australian Army Chaplains Department (RAAChD)', *Oxford Companion to Australian Military History,* 462; Abbott recorded the number as 753, Abbott, *In This Sign Conquer,* 492.
37 There were also 158 ministers who served as welfare officers, including 128 Salvation Army officers.
38 Allen Brook, Letter to FUCCB, 23 July 1947.
39 'AAChD – in *Military History of the Australian Army Chaplains,* Undated MS Typescript, Australian War Memorial 54, 177-2-1 Part 1.
40 Peter Davidson, *Sky Pilot, a History of Chaplaincy in the RAAF 1926–1990* (Canberra: Directorate of Departmental Publications, Department of Defence, 1990), 2-1. Davidson compiled his list from many sources. Apparently no single list of appointments was kept.

(Anglican) was serving full-time,⁴¹ and the remainder – two Anglicans, two Roman Catholics and one Presbyterian – were all part-time members of the Citizens Air Force. Five more were added over the next nine months, and by the end of 1940 the number had increased to fifty-three.⁴² To regularise the appointment of new chaplains an Air Force Chaplaincy Branch was formed on 31 August 1940 with two full-time Staff Chaplains: Ken Morrison (Roman Catholic) and Irving Davidson (Presbyterian).⁴³ They were soon after joined by Joseph Booth (Anglican) and Thomas Rentoul (Methodist). Then, in 1942 Walter Albiston (United Board) joined them.⁴⁴ Eventually the figure reached three hundred and seventy-two chaplains serving full-time and part-time.

Of these twenty-nine were from the United Board which, though less dramatic than the number serving in the Army, was also significantly greater that census figures required. Like those of the major churches, most of the UB chaplains were appointed after Japan entered the war. Nine were Baptists, eight Congregationalists, and six each from Churches of Christ and the Salvation Army. Eight served in Papua New Guinea and the remainder at various bases within Australia.⁴⁵

The Navy's chaplaincy branch, as already mentioned, was by far the smallest of the three services. At the outbreak of war, it only had nine permanent chaplains: seven Anglican, Presbyterian and Methodist, and two Roman Catholics. Over the next six years this increased to thirty full-time chaplains and more than thirty-three reservists.⁴⁶ None of them, however, were UB chaplains.

---

41 Davidson notes that he was the only chaplain listed as a member of the Permanent Air Force, and until 1942 was actually listed as a public servant with the honorary rank of Flight Lieutenant. It was not until 1941 that he appears on the active list of the (CAF).
42 Davidson, *Sky Pilot*, 2-1.
43 'History of Chaplains in the Air Force,' accessed 11 October 2022, https://www.airforce.gov.au/our-people/our-culture/chaplains/history.
44 Status and Advancement of Chaplains in the Time of War, NAA A14487, 22/AB/3722.
45 Davidson, *Sky Pilot*, Statistical Summary RAAF Chaplains – WWII and Roll and Service Record RAAF Chaplains 1926–1990, unnumbered pages.
46 Strong, *Chaplains in the Royal Australian Navy*, 153.

Appointment of a Fifth Chaplain General

Recognition of the United Board as an equal participant with the major churches finally came in 1940 with the appointment of a fifth Chaplain General to represent its member churches. The matter was raised by the Attorney General in a submission to the Military Board on 28 November 1939 advocating the appointment of a Chaplain General representing all Protestant Denominations not covered by existing regulations, which provided only for Chaplain Generals for the Anglican, Roman Catholic, Methodist and Presbyterian Churches. Agendum No. 246/1939 added:

> In administering the Chaplains' Department, there are many questions which arise which necessitate collaboration with Chaplains-General as representing the four largest denominations. As there is no representative Chaplain-General for the other denominations, it is not practicable to deal with the Senior Chaplain of each other denomination in each Military District ... A Chaplain-General representative of all Other Protestant Denominations is required.[47]

The Secretary of the Military Board, in a letter to the Secretary of the Attorney General's Department, agreed with the proposal[48] and the Board Minute recorded that it concurred with the recommendation, and that 'the Other Protestant Denominations for whom the provision of a Chaplain-General is recommended comprise chiefly Baptists, Congregationalists, and Church of Christ'.[49]

On this occasion there was no voiced opposition from the existing Chaplains General, only one of whom, Archbishop Daniel Mannix, had been part of the previous group that had objected to such an appointment in 1915.[50] The Defence Act, 1941 was subsequently amended to read:

---

47 Appointment OPD Chaplain General, NAA MP508/1, 56/701/2, 7.
48 Appointment OPD Chaplain General, NAA MP508/1, 56/701/2, 1-2.
49 Appointment OPD Chaplain General, NAA MP508/1, 56/701/2, 6.
50 He was the only Chaplain General not to have gone on record as opposing the appointment (see Abbott, 409).

Five chaplains-general may be appointed, including one chaplain-general for each of the following denominations: Anglican, Roman Catholic, Presbyterian, Methodist, and one chaplain-general for all such Protestant denominations not mentioned in this regulation as are specified by the Governor-General in the instrument of appointments of the last-mentioned chaplain-general.[51]

The first person appointed to this position was Norman Victor Hansen (Baptist). A report in the *Australian Christian* spoke of the delight this appointment gave to serving United Board chaplains who, at last, felt they had achieved parity with chaplains of the major churches. A gathering at the Northcote Baptist Church on 2 October welcomed and congratulated him on his appointment and presented him with a 'volume suitably inscribed ... as a memento of the occasion'.[52]

Hansen was born in Melbourne in 1892 into a family that belonged to a strict Brethren church. In his late teens he joined the local Baptist Cricket Club, which ultimately led to him becoming a member of the Baptist Church, later enrolling as a student at the Baptist College of Victoria.[53] On 15 June 1915, while still a theological student, he enlisted in the AIF and was posted to the Australian Army Medical Corps serving in Egypt and France. On 1 September 1917 he was posted to the 7th Battalion where he was promoted to corporal on 21 August 1918. He was subsequently sent to England for officer training and was commissioned as a 2nd lieutenant on 5 January 1919, and then to full lieutenant on 5 April 1919. He returned to Australia in July 1919 and was discharged on 18 September.[54]

Hansen returned to his studies at the Baptist College and was eventually ordained in 1921. He then ministered in South Australia, including serving as President of the Baptist Union of South Australia in 1934. In 1937 he returned to Victoria and pastored the Northcote Baptist Church for five years before moving back to South Australia.

---

51 Commonwealth of Australia, Defence Act, 1941, para 1045.
52 *Australian Christian*, 30 October 1940, 631.
53 Abbott, *In this Sign Conquer*, 409-410.
54 Hansen Norman, Military Record, NAA B2455, 1-35.

His appointment as Chaplain General took effect on 22 July 1940.[55] However, it was to be a brief appointment that only lasted until 1942. Abbott describes him as a 'self-effacing man' who 'If his appointment had been *allowed* to continue beyond 1942 ... could have made a significant contribution to the office of Chaplain General'.[56] Abbott does not elaborate on the reason for Hansen's departure but appears to suggest that he may not have had the support of the State United Board committees. He comments that in South Australia the Baptists considered Hansen to be 'a radical, if not a renegade' and was 'liberal in his theology'.[57] Unfortunately, the earliest extant minutes of the United Board and its successor, the United Churches Chaplaincy Board,[58] postdate the time of his departure and make no mention of him. However, if Abbott is correct, Hansen's liberalism may well have cost him the support of his theologically conservative denomination.

Following a three-month interim appointment, when James Thomas (Churches of Christ) served as Acting Chaplain General,[59] he was succeeded by Allen Brooke (Churches of Christ), who was to continue in this role for the next 22 years. Like Hansen, Brooke had served as a soldier in World War I. He enlisted in the 1st AIF just before his seventeenth birthday (having put his age up a year) on 11 January 1916. He embarked from Fremantle on 17 April aboard the transport *Aeneas*, eventually reaching the Front in December, where he was posted to the 27th Battalion and later to the 2nd Division Signal Company. He eventually arrived back in Australia on 5 October 1919 and was discharged on 29 November.[60]

In 1922 he entered the Churches of Christ College of the Bible in Melbourne, graduating three years later. He was ordained as a minister of the Churches of Christ and for the next sixteen years ministered in various congregations in New Zealand, South Australia, Victoria,

---

55 Department of Army [Chaplains] – Senior Chaplain, United Board 3 M.D., NAA MP508/1, 56/701/32.
56 Abbott, *In this Sign Conquer*, 410-411.
57 Abbott, *In This Sign Conquer*, 410.
58 *Victorian State United Churches Chaplaincy Board*, 21 December 1942.
59 'AAChD – History 1939–1946, United Churches,' in *Military History of the Australian Army Chaplains,* undated MS typescript, AWM 54, 177/2/1, Part 1,.
60 Brooke Allen Military Record, NAA B2455.

Western Australia and Queensland, where he also served as Conference President while pastoring the Ann Street Church in Brisbane.

It was from this congregation that he enlisted in the 2nd AIF as a Chaplain 4th Class on 1 August 1940 and was posted to Headquarters 1 Australian Corps. He embarked for England with the 2/1 Machine Gun Regiment aboard the transport *Ulysses,* arriving on 9 October, and was posted to Headquarters 25 Infantry Brigade. He deployed with them to the Middle East three months later, arriving on 8 March 1941, and on 2 June was detached for service with the 2/23 Battalion, which was then part of the garrison of Tobruk. A period of illness caused him to be medically evacuated to Australia, arriving on 26 August. He was admitted to 112 Australian General Hospital in Brisbane and remained there until 19 March 1942 when he was reclassified as medically unfit for service with the AIF. On 14 September he was promoted to Chaplain 1st Class then appointed Chaplain General (United Board) at Land Headquarters in Melbourne.[61]

Brooke was not the only nominee for the position. William Crossman (Churches of Christ), who likewise held the rank of Chaplain 1st Class, was also nominated by the United Board in New South Wales. It is curious that Brooke, who had only been a chaplain for two years and had had a meteoric rise to Chaplain 1st Class and then to Chaplain General, should have been nominated rather than Crossman, whose chaplaincy experience was so much greater. Years later, his son, Chaplain Air Commodore Geoff Crossman RAAF Ret'd. reflected:

> I remember the story from my late Dad. Nominations for the first Chaplain General (United Churches) were called; each State Board could nominate. My Dad was nominated from New South Wales (Eastern Command), and Allen Brook from Queensland (Northern Command) ... in order not to split the vote denominationally, Dad withdrew and left Allen the field, as the only Churches of Christ nomination.[62]

This seems to have been typical of Crossman who characteristically never

---

61 Brooke Allen Military Record, NAA B2455.
62 G.J. Crossman, letter to Douglas Abbott, 10 April 1989; Abbott, *In This Sign Conquer,* 413.

put his own interests above those of the Gospel. Dennis Nutt said of him: 'The keynote of his life and ministry was his loyalty to Christ and the plea for the restoration of New Testament Christianity in its fruits as well as in its doctrines and ordinances'. Evidence of the esteem in which he was held is seen in the number of prominent leaders of Churches of Christ who, on 12 October 1958, participated in his funeral; 'a veritable *Who's Who* of the senior ministers in the State'.[63]

Brooke's appointment as successor to Hansen marked the beginning of a series of rotations with Churches of Christ nominees succeeding Baptists and viceversa until the Conference of Chaplains General was replaced by the Religious Advisory Committee to the Services in 1982. It then continued in the new body until 2019 when Ralph Estherby, a minister of the Australian Christian Churches, became the nominee. Brooke served full-time for the duration of the war and then returned to the normal part-time arrangement, serving continuously until his retirement in 1964. This included fifteen years as secretary of the Conference of Chaplains General, a role he performed with great professional competence.

Abbott describes Brooke's military career as 'unique and outstanding in the annals of the Australian Army'. He refers to him 'as the "face" of army chaplaincy at Army Headquarters', being held in high regard by many senior officers and senior chaplains. He is, however, also critical of Brooke as being 'essentially a *status quo* man' who liked 'to play the general' and referred to himself as 'General Brooke, whilst at the same time making every possible provision that other chaplains would ... not [be allowed] to be rank or status conscious'.[64] It is difficult from this distance to judge whether this criticism is fair or not. Abbott's history of the Chaplains General tends to be critical of them throughout, and he himself had a reputation among fellow chaplains of being quite opposed to them.[65] Even so, his assessment of Brooke's professional competence and accomplishments is complimentary, as was Brooke's reputation within the United Board, evidenced by the public tribute paid to him on his retirement.

---

63 Dennis Nutt, 'A Life Well Lived', *Churches of Christ Historical Digest*, Issue 181, November 2013.
64 Abbott, *In This Sign Conquer*, 414.
65 Author's reminiscences of having served with Abbott.

The creation of the fifth Chaplain General position coincided with a long-overdue overhaul of Army chaplaincy command and staff structures which, as Gladwin comments, reflected the 'wider reforms of the Australian Army as a whole ... implemented in the wake of General Blamey's appointment as Commander-in-Chief of the Australian Military Forces'.[66] The effectiveness of the Chaplains General was greatly enhanced by making them directly answerable to the Army's Adjutant-General branch through their own Deputy Assistant Adjutant-General, and by providing them with dedicated staff. In addition, whereas the four Chaplains General had previously worked individually and on an *ad hoc* basis, from June 1942 they, including the newly appointed United Board member, began to meet regularly in conference. This enabled them to address the larger issues of policy in addition to their traditional denominational responsibilities. The reorganisation became complete when the Chaplains Department Headquarters moved into its own premises at 477 St Kilda Rd, South Melbourne.[67]

One of the most important consequences of this, especially for the United Board, related to the increased tempo of mobilization that followed Japan's entry into the war and the immediate threat to Australia. As the size of the Army grew so did the number of chaplaincy positions. A new war establishment was created in July 1942[68] which, as Tom Johnstone notes, required:

> Chaplaincy services for an increased number of field force headquarters ... a vastly increased number of combat units ... [and] for chaplaincy services within Australia ... across vast lines of communications, many logistic bases and numerous training establishments.[69]

This expanded establishment produced a new set of cracks in the wall built on the principle of proportionate representation. The major churches, especially the Anglicans, found themselves under stress to

---

66 Gladwin, *Captains of the Soul*, 163.
67 Abbott, *In This Sign Conquer*, 60.
68 Australian Army's Chaplain Department – Organisation, NAA MP508/1, 96/707945.
69 Tom Johnstone, *The Cross of Anzac, Australian Catholic Service Chaplains* (Brisbane: Church Archivists' Press, 2003), 91.

produce the required number of acceptable candidates for existing vacancies. But now, with its own Chaplain General sitting as an equal partner among the other Chaplains General, the United Board was able to offer the conference a wider pool of potential applicants to fill the vacant positions. The Chaplains General euphemistically referred to this as 'horse-trading' and such appointments were seen as temporary until official ratios could be restored. But in reality, they continued to expand until demobilisation reduced chaplaincy numbers to peace-time levels.

Hansen and Brooke, especially the latter, proved themselves to be valuable assets to the Conference of Chaplains General. They were unique among that group in that both of them had served as ordinary soldiers during World War I, and their familiarity with the life of soldiers on active service was a helpful foundation for understanding the qualities required of effective army chaplains. Brooke's experience as a unit chaplain added to his capability, even though it had been relatively short. Of the other seven Chaplains General who served during World War II, five had been unit chaplains during World War I and two had no previous military experience. But none could quite match the breadth of operational experience that Brooke and Hansen brought to the role.[70]

The Appointment of a Fifth Air Force Staff Chaplain

A similar boost to the status of the United Board took place in 1942 when Walter Albiston (Congregationalist) was elevated to the position of Staff Chaplain, representing what was then transitioning from being known as the United Board to the United Churches.[71] He joined the four full-time Staff Chaplains as the only part-time member of that team, and continued in it until 1952 when he was appointed Principal Chaplain.[72] He was the first United Board/United Churches chaplain to be appointed to the Air Force and was originally posted to RAAF Ascot

---

70 Abbott, *In This Sign Conquer*, 304 ff.
71 The actual date is indeterminate due to the lack of relevant minutes. Hence, the designation United Board/United Churches will be used for this transitional period.
72 Davidson, *Sky Pilot*, 6-2, Statistical Summary RAAF Chaplains.

Vale in late 1940. His advancement to Staff Chaplain a mere two years later is an indication of the growing recognition of the United Board/ United Churches, which by then had ten chaplains serving, including three full-time.[73]

Albiston was already a formidable figure within the Congregational Union. He was ordained in 1914, and ministered at Warrnambool where he earned a reputation for vigorously supporting conscription and loyalty to the British Empire. In 1918 he moved to Ballarat and founded the Victorian Protestant Federation, where he issued a manifesto opposing the policies of Roman Catholic Archbishop Mannix. He was also a founder of the monthly magazine *Vigilant*, through which he campaigned against government funding of religious schools, hospitals and alleged Roman Catholic lobbying of government. In October 1925 he was elected Secretary of the Congregational Union of Victoria, and in 1941 became Secretary of the Congregational Union of Australia and New Zealand. He was a powerful preacher, an excellent administrator and a convinced ecumenist. He brought these gifts to his role as Staff Chaplain and used them effectively as part of the team that administered Air Force chaplaincy during its first and greatest challenge.[74] Under his leadership the United Board/United Churches became an established partner in the Air Force Chaplains Branch, as it had in the Army Chaplains Department.

Lutheran Church

The third significant factor in the re-shaping of United Board chaplaincy was the inclusion of chaplains from the two branches of the Lutheran Church: the Evangelical Lutheran Synod of Australia (ELSA)[75] and the United Evangelical Lutheran Church of Australia (UELCA). Five Lutheran ministers were commissioned as Australian Army chaplains during World War II, three serving full-time. Their appointment was of particular importance to Australian Lutherans in that it demonstrated

---

73 Davidson, *Sky Pilot*, Roll and Service Record RAAF Chaplains 1926–1990, unnumbered pages.
74 Niel Gunson, 'Albiston, Walter (1889-1965),' *ADB*, Volume 7 (Melbourne: MUP, 1979), 29.
75 Which in 1944 became the Evangelical Lutheran Church of Australia.

to the wider community that they were loyal Australians and not, as often believed, German sympathisers.

To understand the importance of this it is necessary to recognise the level of antipathy towards Lutherans that had existed for at least two generations. Breward refers to the high price paid by Lutherans in World War I 'who, even when family members were serving overseas [in the 1st AIF], found themselves accused of being traitors and supporters of Germany', and how the '1915 Commonwealth War Precautions Act gave the federal government power to intern without trial or reasons given'.[76] But by 1939, and the declaration of another war against Germany, the nation was ready to accept Lutheran chaplains into its army and the Lutheran Churches' declaration of loyalty. UELCA President, Dr Johannes Stolz, wrote to the Governor General, Lord Gowrie, pledging 'unswerving loyalty to the King and Government,' assuring him that Australia and the British Empire's cause was also that of Australian Lutherans, and asking for protection against the sort of abuse that Lutherans experienced during World War I.[77]

The first Lutheran chaplains were appointed in 1940. On 5 September the Military Secretary informed Hansen that the Adjutant-General had approved the appointment of one UELCA chaplain in Eastern Command and one in 4th Military District.[78] Over the next couple of years several Lutheran pastors were nominated by the Presidents of the UELCA (Dr Johannes Stolz) and the ELSA (Pastor Clemens Hoopman),[79] and on 8 June 1943 the Deputy Adjutant-General reported that Chaplains Herbert Winkler (ELSA) and Werner Petering (UELCA) were both on full-time duty.[80] Three months later, on 10 September, Brooke reported that Chaplains Mervyn Stolz (UELCA) and Otto Thiele (ELSA) had been placed on the Unattached List pending posting.[81]

As the mobilisation of troops began and the training camps were established, the General Council of the UELCA appointed Pastor

---

76 Breward, *A History of the Australian Churches*, 108.
77 Everard Leske, *The Story of Lutherans and Lutheranism in Australia, 1838–1996* (Adelaide: Friends of Lutheran Archives, 2009), 160.
78 Appointment of Lutheran Chaplains to Army, NAA MP742/1, 56/1/17, 7.
79 Appointment of Lutheran Chaplains to Army, NAA MP742/1, 56/1/17, 13.
80 Appointment of Lutheran Chaplains to Army, NAA MP742/1, 56/1/17, 5.
81 Appointment of Lutheran Chaplains to Army, NAA MP742/1, 56/1/17, 1-2.

Christoph Stolz to keep in touch with Lutherans serving in the AIF. As the war progressed the General Council expanded this work. On 26 March 1942 the President General reported that an auxiliary for welfare work had been established in each State where the UELCA was present 'for the spiritual care of members of our Church serving with the fighting forces'.[82] Pastor Stolz, however, unlike his namesake Mervyn Stolz, was not a commissioned army chaplain.

Chaplain Herbert Winkler, who enlisted on 24 December 1942, reveals the evangelical conservatism typical of Lutheran chaplains thus:

> As the first Lutheran Chaplain [ELSA] to serve with the AIF, with Army ways to learn, and the need to prove that despite rigid limits of doctrine and conscience a Lutheran chaplain could and would serve the needs of all Christian men, my task, perhaps, bristled with added difficulties ... by far the majority [of chaplains] impressed me as earnest and sincere in their desire to live and serve as godly men ... for a small minority ... I too think their conduct indefensible ... these foolish fellows thought they would attract men to their "services" in which they either soft-pedalled on Christian doctrine, or failed to speak at all of the Word and Will of God.[83]

The commissioning of these men launched a fruitful relationship between what is now the Lutheran Church of Australia and the renamed Federal United Churches Chaplaincy Board.

The Federal United Churches Chaplaincy Board

The final step in the re-shaping of what began as the Other Protestant Denominations was the transition in 1942 from State based United Chaplaincy Boards to a federally based board known as The Federal United Churches Chaplaincy Board (FUCCB). The State boards continued to work in cooperation with it, and chaplains belonging to

---

82 J.J. Stolz, Address to UELCA, 26 March 1942, Lutheran Church of Australia Archives, Adelaide.
83 M.H. Winkler, 'A Padre Speaks,' in *On Service with the Men and Women of the Evangelical Lutheran Church* (Adelaide: The Service Commission of the Evangelical Lutheran Church of Australia, undated), 41-44.

its member churches would henceforth be referred to as belonging to United Churches (UC).[84]

As previously mentioned, the minutes of FUCCB meetings held prior to 1958 have been lost, but those of the meeting of the 'newly constituted' Victorian 'United State Board'[85] held 21 December 1942 give a good picture of what was happening federally. They record a report from Brooke that the United Churches at that time had sixty-four chaplains on duty. They also included a copy of the constitution used by all State Boards, which clearly would have reflected the Federal Board's constitution. It lists the constituents as the Baptist, Congregational, Churches of Christ and Salvation Army denominations within that State.[86] The members were to include representatives from each denomination (including a serving chaplain, if possible), appointed by the Denominational Executive, together with the Senior Chaplain elected by the Board in consultation with the Chaplain General. Their duties were:

> a) To receive nominations for chaplains for the services (Army, Navy and Air Force) from the relevant Denominational Executives; b) To confer with Senior Chaplain (Army), Staff Chaplain (Air Force) and Chief Chaplain (Navy – if and when appointed)[87] in recommending nominees for appointment as chaplains; c) To receive and administer funds for chaplaincy work; d) To support, in every way, all State United Board Chaplains on service; e) To cooperate with the Federal Board in all matters pertaining to chaplaincies.[88]

The creation of the Federal Board finally brought this 'aggregate of quite non-coherent and non-corporate factors having no solidarity or unified organic administration' into line with the chaplaincy administration of the major churches. Its reputation, already well established by the end

---

84 However, it took years for the earlier designations to die out in common usage.
85 It appears that the State Boards continued (initially at least) to refer to themselves as 'United Chaplaincy Boards'.
86 The Lutheran Church of Australia did not officially become a member of FUCCB until 1962, even though its nominated chaplains were associated with FUCCB; Nigel Long, email message to author, 28 March 2022.
87 Clearly, there was still a hope that United Churches chaplains might be appointed to Navy.
88 Victorian United State Board, *Minutes of Meeting*, 21 December 1942.

of World War I, survived the two decades of inactivity and neglect that characterised the armed services between the wars, and ensured its place with the major churches in the rapid expansion of military chaplaincy during World War II. The appointment of its own dedicated Chaplain General in 1940 and Air Force Staff Chaplain in 1942, working with the newly formed Federal Board, provided the framework for its greatest achievement, the commissioning of one hundred and seventy-three chaplains for the Army and twenty-nine for the Air Force, significantly more than that required under the policy of proportionality. It was made possible because the United Churches were now represented at the highest level of Army and Air Force chaplaincy governance, able to provide new chaplains to fill the gaps caused by shortages from the major churches. If one adds the one hundred and fifty-eight philanthropic representatives, most of whom were ministers of Federal Board churches,[89] the achievement is even more remarkable. Based on proportionate representation, FUCCB was required to provide five percent of the total chaplaincy numbers. However, when Australia faced its greatest crisis, the once rejected and now reshaped stone rose to the challenge and, proportionately, surpassed all the others.

---

89 Who were often thought of as 'padres' even though they were not commissioned chaplains.

CHAPTER SIX

World War II: The Re-weathering of the Stone

A New Generation of Chaplains

Peter Fitzsimons, commenting on the raising of the 2nd Australian Imperial Force as a worthy successor of the 1st AIF, records that the universal aspiration of the young men who enlisted was that they would 'be as good and as worthy of deep respect as their famous fathers'.[1] The original Anzacs, initially dismissed as ill-disciplined and of doubtful value, created a reputation that was second to none among the allied armies of World War I.[2] Within Australia, that reputation took on legendary status during the inter-war decades,[3] immortalized by the poet John Masefield's description of them as: 'The finest young men I have seen'.[4] It is little wonder that the thousands who flocked to enlist felt themselves under a sacred obligation to emulate those who had gone before. A post-war survey revealed a sense of duty as the single most important factor that motivated their enlistment.[5]

This was equally true of their chaplains. As already noted, of the

---

1 Peter Fitzsimons, *Tobruk,* (Sydney: Harper Collins, 2006), 51.
2 Charles Bean, *Anzac to Amiens* (Canberra, Australian War Memorial, 1983), 536-537.
3 Bean, *Anzac to Amiens,* 181.
4 *Recorder,* 18 October 1934, 4; John Vader, 'The Anzacs,' in *History of the First World War,* Volume 3, edited by Barrie Pitt (London: Purnell, 1970), 1044-1046.
5 Mark Johnston, *Anzacs in the Middle East* (Melbourne: Cambridge University Press, 2013), 2.

seven hundred and fifty-four Army chaplains who served full-time in World War II, one hundred and four were from the United Board/United Churches – almost four times as many as their OPD predecessors in World War I. It was a further indication of how the contribution of this numerically insignificant group of churches was considerably greater than mere census figures required.[6] This chapter demonstrates its growth from the strictly proportionate number deployed in the early years of the war to its significantly disproportionate contribution from 1942–1945.

Chaplaincy Training

Whereas most of the chaplains of the 1st AIF entered the Army with little idea of how they were to translate their ministry expertise into this strange and intimidating world, the Chaplains Department in World War II attempted to provide an orientation.[7] Responding to pressure from chaplains returning from the Middle East, the Chaplains Department began to organise chaplains' schools early in the war against Japan.[8] Before this the presence of World War I veterans, whose experience and reputation had been forged in the most arduous conditions imaginable, provided expertise of incalculable value to a new generation of chaplains. One such was the previously mentioned Arthur Forbes (Churches of Christ), who continued to serve during the inter-war period and volunteered for full-time service again during World War II.[9]

The first attempt to tap this accumulated wealth of experience came in 1939 when Anglican Chaplain General, Henry Le Fanu, issued instructional guidelines on the conduct of services, and personal dealings with officers, men and other chaplains. His successor, Charles Riley, Bishop of Bendigo, followed with a series of leaflets in 1942.

---

6   Appointment of Chaplains, NAA MP508/1, 56/570/340.
7   Victorian State United Churches Chaplaincy Board, *Minutes of Meeting*, 25 January 1943.
8   R.W. Tippett, 'Australian Army Chaplains, South West Pacific Area, 1942–1945,' Unpublished Master's thesis, University of New South Wales, 1989, 83; Douglas Abbott, *In This Sign Conquer: The Chaplains General of the Australian Army, 1913–1981,'* unpublished manuscript, 1995, 179.
9   Arthur Bottrell, 'Forbes, Arthur Edward (1881–1946)', *ADB*, Vol. 8 (Melbourne: MUP, 1981), 539.

These proved so popular that the other Chaplains General requested copies to circulate among their chaplains.[10] Riley had previously issued a helpful paper on Army Chaplaincy written by Rev Kenneth Henderson, an Anglican who had served as chaplain for the 12th Brigade during World War I.[11]

Henderson, with assistance from other veteran chaplains, produced what he called 'Brief notes of guidance and advice with a view to helping newly appointed chaplains to achieve their maximum efficiency as soon as possible'. He highlighted the difficulty that all new chaplains face in having no one to advise them as to what they were required to do, these being 'rightly regarded as belonging to his own personality and training', leaving them entirely dependent on their own initiative and resources in an environment bearing little resemblance to that of the local church. He listed the various methods World War I chaplains used to 'break the ice' and make meaningful contact with men, most of whom had little or no contact with the church. He advised them to experiment until they discovered what would 'quicken the interest of some men'. He warned against 'the intoxication of the uniform,' and reminded them that they were serving as ministers of religion not as military officers. He urged them to avoid 'an exaggerated "heartiness" of manner' but to seek to overcome prejudice by being faithful to their work 'in an unostentatious friendly kind of way'. He gave helpful advice on the conduct of parades, visitation of the sick, dealing with grievances and, most importantly, where to locate themselves for best effect during the various phases of military operations. He also reminded them not to be discouraged if their work seemed to lack visible result. The real results, he affirmed, 'Are produced in the inner disposition of men to the values and conduct that you stand for'.[12]

Another contribution to this body of shared experience came from Allen Brooke. Two years before his appointment as Chaplain General, and while travelling to the UK with his unit, he reflected on the

---

10 Tippett, 'Australian Army Chaplains,' 43.
11 Henderson K.T. Military Record, NAA B2455, 5338747.
12 Kenneth Thorne Henderson, 'Memorandum issued for the personal use of C. of E. Chaplains serving in the A.I.F,' (Bendigo: Deputy Anglican Chaplain General); Tippett, 'Australian Army Chaplains,' Appendix 1, 328-330.

qualities 'demanded of a padre seriously concerned with doing his job'. He listed *faithfulness* as being first and foremost: faithfulness to God as 'the great Commander, and faithfulness to his comrades' which, echoing the practical ecumenism of chaplains in World War I, would reveal itself in a determination 'to lose sight of all religious distinctions and strive to be faithful to every man for the sake of Christ'. The second characteristic he advocated was *friendliness*. He referred to the Good Shepherd of the Bible who 'knows his sheep individually and spends time caring for their welfare', insisting that 'the padre tries to be that kind of pastor'. The third quality was *fatherliness*. Commenting that the term 'padre' literally means 'father' he stressed that a good chaplain, who often is significantly older than the troops he serves, should be 'a father to those in his spiritual care' and 'cultivate a fatherly affection that smiles, and suffers, and sympathises, and serves'.[13]

The War in the Middle East

The first chaplains to put this advice to the test were those who served in the Middle Eastern campaigns of 1940–1942. From September 1940 to July 1941, the 2nd AIF was continuously in action against Italian and German troops in defence of Egypt and the Suez Canal. Rodney Tippett and Douglas Abbott, drawing upon what they describe as incomplete records held by the Army Chaplains Department, but which have now been lost,[14] counted approximately seventy chaplains deployed to those operations.[15] At least five were UB chaplains: Robert Helmore[16] and James Salter[17] (Baptists), Ernest Miles[18] and Allen Brooke[19] (Churches of Christ), and Charlie Watts[20] (Congregational). Arthur McIlveen (Salvation Army), though not officially a chaplain but rather a Salvation Army philanthropic representative, also served as chaplain to the 2/9th

---

13 *Australian Christian*, 23 October 1940, 610-611.
14 Darren Jaensch, email to Robert Smith, 1 December 2022.
15 Tippett, 'Australian Army Chaplains,' 28; Abbott, 'In This Sign Conquer,' 492-501.
16 Helmore R.A Military Record, NAA B883, VX197.
17 Salter J.C. Military Record, NAA B883, TX6008.
18 Miles E.J. Military Record, NAA B883, WX11128.
19 Brooke Allen Military Record, NAA B2455.
20 Watts C. Military Record, NAA B883, SX8186.

Battalion, having been directly appointed battalion chaplain by Sir Leslie Morshead, Commander of the Australian 9th Division.[21]

Commenting on the 9th Division's celebrated role in the defence of Tobruk during 1941–1942, Michael Gladwin refers to the 'crucial and largely unsung role that chaplains played, and how their quiet contribution was nowhere better exemplified than in the efforts of McIlveen, or 'Padre Mac' as the troops called him'.[22] Four UB chaplains – Salter, Miles, Brooke and McIlveen – were present during the siege of Tobruk, but McIlveen is worthy of special attention, not only because the manner of his appointment was unique, but also because he became to the 2nd AIF what his fellow Salvationist, William McKenzie, had been to the 1st AIF. Nelson Dunster records that at the outbreak of World War II McIlveen was a 53-year-old Salvation Army officer who had been serving as a CMF chaplain since 1926. He considered that his age would preclude him from service as a chaplain with the AIF[23] so he obtained an appointment as the Salvation Army's first Red Shield overseas representative.

He sailed to war with the 18th Brigade on 5 May 1940 aboard the transport *Mauretania*, arriving in Scotland on 16 June, whence he travelled to the training area on Salisbury Plain. It was there that Major General Morshead, impressed with McIlveen's service to the troops and aware of the need for a battalion chaplain, appointed him unofficial chaplain to the 2/9th Battalion.[24] Morshead did this without reference to the Chaplains General and with disregard for the policies that tied chaplaincy appointments to clergy nominated by approved denominations through the relevant Chaplain General, or denominational authority in the case of the United Board.[25] This was contrary to the principle that chaplains are essentially ministers of the Church on loan to the military, and though subject to military command and control

---

21 Darryl McIntyre, 'McIlveen, Sir Arthur William (1886–1979),' *ADB*, Volume 15 (Melbourne: MUP, 2000); Michael Gladwin, *Captains of the Soul. A History of Australian Army Chaplains* (Sydney: Big Sky), 111; McIlveen Arthur, AWM, REL.30722.002.
22 Gladwin, *Captains of the Soul*, 110.
23 McIlveen AW. Military Record, NAA B2455.
24 Nelson Dunster, *Padre to the 'Rats'* (London: Salvationist Publishing and Supplies,1971), 66.
25 McIlveen's appointment was made shortly before the appointment of the United Board's first Chaplain General.

in a general sense, as ministers of the Church their technical control is through their Chaplain General. In McIlveen's case, this irregularity was a fortunate disregarding of the rules.

He sailed from Glasgow with the 18th Brigade on 17 November aboard the *Strathaird*, disembarking in Alexandria on 31 December. His brigade then travelled west to Tobruk and became part of the garrison defending it against repeated attacks by Rommel's Axis forces. It was there that McIlveen's reputation was fully established. Dunster quotes one of the 'Rats of Tobruk,' Major William Isaacs, who described McIlveen as 'The best-liked member of the garrison' and said of him:

> In rest areas there was never any need for a compulsory church parade when it became known that Padre Mac would be doing the honours. At least one battalion used to get 100 percent attendances ... It is hard to define just how Padre Mac helped us – he was so unobtrusive ... whenever a patrol would come back ... he was always there with a cheery word and a kerosine tin of coffee. He travelled tremendous distances on foot every day, bringing cheer wherever he went.[26]

McIlveen had what one ex-soldier called a 'secret weapon': an old portable, handle-turned gramophone that he took with him and played wherever he went.[27] It became so famous that it is now on permanent display in the Australian War Memorial.[28] The *Newcastle Morning Herald and Miners' Advocate* said of him:

> Brigadier[29] McIlveen was known during the Siege of Tobruk as 'the man with the gramophone'. [He] left Australia by troopship in May 1940 taking with him a second-hand portable gramophone and some records. He played these on the ship, in camps, hospitals and air-raid shelters in England and later at Tobruk. The records were in constant demand and helped build the morale of men who were fighting homesickness and

---

26 Dunster, *Padre to the Rats,* 70.
27 *Diggers Digest,* 9 July 1948.
28 McIlveen Arthur, AWM REL.30722.002.
29 His Salvation Army Rank after World War II.

despondency. Records damaged by shell blast were patched together with sticking plaster and made almost as good as new.[30]

'Padre Mac' was a worthy successor to the unforgettable 'Fighting Mac' of World War I and helped build the enormous goodwill that Australian ex-servicemen have for the Salvation Army. Unofficial though his chaplaincy status was, and unable though he was to celebrate the sacraments, his preaching, pastoral ministry and practical support to soldiers brought great credit to United Board chaplaincy.[31]

There is evidence that the Salvation Army did not always meet with the approval of other chaplains. Tippett refers to the reputation that Salvation Army officers built in World War I, by their mixture of religion and practical help, becoming 'the measure by which others [chaplains] were judged'. However, the reorganisation of the Army Chaplains Department and the 'growing recognition that there was a difference between chaplaincy and welfare work' caused many of the Salvationists (like McIlveen) to became welfare officers, for which they had special training and experience.[32] The twelve who remained as full-time chaplains, and twenty-eight who served part-time, were once again absorbed within the United Board/United Churches group.

The occasional strained relationships with other chaplains, probably Anglican and Roman Catholic, appear to be related to a perception that the Salvation Army wanted to have the best of both worlds, claiming to be a welfare organisation in order to share the proceeds of the Lord Mayor's Comfort Fund, while also claiming to be a denomination entitled to nominate chaplains. In contrast, the 'Anglicans and Roman Catholics entirely funded their own very considerable welfare work'.[33]

It is unlikely that United Board/United Churches chaplains shared these concerns about the Salvation Army. They relied heavily on the welfare work of the Red Shield Services, as is evident in the Victorian State United Churches Chaplaincy Board's appreciative reference to 'the vast amount of work being done by the Salvation Army in the Red

30 *Newcastle Morning Herald and Miners' Advocate*, 24 May 1952, 5.
31 The Salvation Army recalled McIlveen to Australia early in 1942.
32 David Woodbury, 'When the World Went to War, the Salvation Army Was There,' *Halleluiah*, Volume 1, Issue 3, Autumn, 2008, 25.
33 Keith Pither and Ralph Ogden as cited in Tippett, 'Australian Army Chaplains,' 54-55.

Shield huts'. It was this example that provoked that State Board to establish a State Chaplains' Fund to provide its chaplains with a minimum of one hundred pounds each for use in welfare work.[34]

Tensions were also raised when some Red Shield representatives began to conduct religious services, which were the sole responsibility of chaplains. A direction by the Secretary of the Army (February 1942) actually forbade Salvation Army welfare workers from acting as chaplains except when no chaplain was available, 'and only then with the sanction of the Senior Chaplain'.[35] Nevertheless, for the troops on the ground there was no difference; they were all padres, and those from the Salvation Army were highly esteemed.

The Navy also played a significant role in this theatre of the war. Ships of the RAN patrolled the sea lanes and supported the campaigns in the Mediterranean,[36] including the sinking of the Italian cruiser *Bartolomeo Colleoni* by HMAS *Sydney*. Anglican, Roman Catholic and Protestant chaplains were part of those ships' companies, but none were drawn from the United Board. The same was true of the Air Force which had three fighter squadrons in the Middle East. Nor were there any United Board/United Churches chaplains among the twelve deployed by the RAAF to the United Kingdom.[37]

By December 1942 the Middle East Campaign was almost won.[38] The number of UB chaplains in that theatre, though relatively small, was a precursor to their later contribution in the Southwest Pacific, which far surpassed expectations based on denominational size. On 7 December, the Japanese attack on Pearl Harbor thrust Australia into its greatest crisis: one of national survival.[39] It was then that chaplaincy in general, and United Board/United Churches chaplaincy in particular, began to expand to a size never seen before or since.

---

34 VSUCCB, *Minutes of Meeting*, 7 August 1944.
35 Clarification of Chaplains or Welfare Workers (Salvation Army Personnel), NAA MP508/1, 245/708/58; Tippett, *Australian Army Chaplains*, 82.
36 Mike Carlton, *Flagship* (Sydney: Random House, 2016), 162-166.
37 Peter Davidson, *Sky Pilot, a History of Chaplaincy in the RAAF 1926–1990* (Canberra: Directorate of Departmental Publications, Department of Defence), 5-1.
38 It officially ended in May 1943 when German and Italian forces finally surrendered to General Montgomery in Tunis: Richard Overy, *World War 11, The Definitive Visual History* (London: Welbeck, 2020), 148-151.
39 Manning Clark, *A Short History of Australia* (Melbourne: Penguin, 2006), 283, 286.

The Home Front

Of the total number of Army chaplains who served in World War II, fifty were veterans of World War I.[40] Most of these, like Arthur Forbes, were considered too old to serve overseas and spent the war either ministering in military camps scattered across Australia, or as chaplains on troop transports and hospital ships. One such was John Ridley, the well-known Baptist evangelist who, in WWI, had won the Military Cross. On 11 May 1943 he was seconded from the CMF for service with the AIF and posted to an engineer unit in North Queensland, then to 2/2 Hospital Ship, including a voyage to the Middle East in late 1943–1944.[41]

For these chaplains the Home Front provided abundant opportunities for ministry. Their activities were for the most part similar to those of their brethren overseas: conducting worship, leading Bible studies, visiting the sick, providing pastoral care and, as Brooke urged, being a friend and father-figure to all. The difference was the safer environment in which they carried out their duties, away from the constant threat of imminent death or wounds. There was, however, one particularly traumatic activity that only chaplains on the home front were required to perform in person. It was the notification to next-of-kin of fatalities in combat and training activities. Of all the tasks they undertook this was probably the most harrowing.

John Kenneth Martin (Churches of Christ) provides an insight into the ecumenical and practical *modus operandi* of those who ministered in the training camps:

> Fellowship with other padres and men in the united church parades, the meeting of other men around a common table, foretold of a future basis being laid for Christian unity, the Sunday evening meeting around the piano with the great hymn writers while the men laid down their pens and games to listen, sing, and pray together, and the quiet hours on Friday evening were times of blessedness ... In the course of my work I visited men who were laid aside in hospital. An opportunity presented

---

40 Gladwin, *Captains of the Soul*, 103.
41 Ridley J.G. Military Record, NAA B883, NX167498, 3-5, 43.

itself to serve them by obtaining some little comforts which they needed.⁴²

Another area of concern for home front chaplains was the moral threat to off-duty servicemen provided by vendors of illicit alcohol, public houses and brothels. In 1940 several newspaper reports indicated that a majority of military chaplains supported wet canteens in military camps as the only effective way of stemming the drift of young men to establishments where their behaviour could not be regulated. In February, *The Age* reported:

> A statement issued by military padres in Ingleburn camp today was that excessive drinking by men on leave was a serious problem. The true and proper corrective was the introduction of wet canteens, they added. Men who volunteered to serve their country at real personal loss and sacrifice should not be deprived of a privilege which was enjoyed to the full by civilians.⁴³

The *Australian Christian* immediately picked up the story including a response from UB Senior Chaplain William Crossman. He was disappointed with the action taken by the Ingleburn chaplains, noting that 'Among the signatures was that of Padre Helmore, United Board, representing Baptists, Congregationalists and Churches of Christ'. Crossman insisted that Helmore, who came from interstate, had no authority to speak for the NSW United Board, which 'had not expressed itself on the subject of wet canteens'. He went on to state emphatically that he was in favour of dry canteens and that churches and temperance committees, in seeking 'to remove this trinity of evil – drink, gambling and lust' should 'advocate for the placing of hotels out of bounds for all men in uniform'.⁴⁴

This matter exposes a surprisingly rare incident of controversy within the United Board, which had no hierarchical structure that determined policy, relying rather on consensus. Crossman's attitude was probably a good reflection of the general feeling within its member churches. The Congregational Union's Assembly had already expressed its support of

---

42 *Australian Christian*, 17 July 1940, 446.
43 *The Age*, 16 February 1940, 8.
44 *Australian Christian*, 21 February 1940, 114, 120.

'the Dry Canteen policy of the Federal Government' and urged that 'the policy be extended to cover all ranks and close all hotels in the close vicinity of camps'.[45] Alcohol abuse was a primary focus of Church bodies, and the Women's Christian Temperance Union had developed a powerful network across the nation.[46] Graeme Chapman describes the Churches of Christ as being 'wholly opposed to the liquor interests',[47] which was also true of the Baptists. Furthermore, the Salvation Army required its members to be total abstainers.

In addition to this, memories of the mutiny that took place at the Casula camp in February 1916 were still alive. Objecting to an increase in daily training, around fifteen thousand mutineers went on a rampage, stealing large quantities of liquor from hotels, breaking into shops and maltreating pedestrians.[48] The *Daily Telegraph* quoted the President of the Baptist Union and other Protestant leaders who claimed that 'the root of all the trouble was the liquor evil, and ... insisted that ... all hotels near military camps should be closed permanently'.[49] It is little wonder that people like Crossman feared history repeating itself.

The controversy over wet canteens continued to rage and newspapers like the *Armidale Express* were still reporting it months later.[50] It was not until April 1942 that the matter was finally settled when the Minister for the Army announced that wet canteens were to be established in military camps throughout Australia 'as rapidly as possible'.[51] This decision, while vindicating the advocacy of younger chaplains like Helmore, went against the strongly held temperance principles of their churches. It reveals the tension that sometimes exists between the pragmatism of those ministering on the ground and the idealism of those who view things from a distance. It appears that older chaplains, like Crossman, did not support the concept of harm minimisation, whereas most of the younger chaplains did.

45 Congregational Union of New South Wales Yearbook 1940, 39.
46 Ian Breward, *A History of the Australian Christian Churches* (Sydney: Allen & Unwin, 1993), 87, 123.
47 Graeme Chapman, *One Lord, One Faith, One Baptism* (Melbourne: Vital, 1979), 117.
48 *Sydney Morning Herald*, 14 February 1916, 9.
49 *Daily Telegraph*, 16 February 1916, 8,9; F.K. Crowley, *Modern Australia in Documents, 1901–1939* (Melbourne: Wren, 1973), 253.
50 *Armidale Express and New England General Advertiser*, 23 September 1940, 1.
51 *Burrowa News*, 17 April 1942, 1.

Concerns about the moral threats to off-duty servicemen also led to a more practical project in the provision of hostels for servicemen. These facilities were designed to cater for servicemen on leave who, for whatever reason, were not able to go to their own homes, or were on their way to them. They were established to provide an alternative to the hotels and other establishments where young men could easily get into trouble. In Melbourne the Baptist Union's Goble Hall in Albert St. and the Gospel Hall in Collins St. both provided meals, sleeping accommodation and other services for service personnel. By 1944 it was reported that United Churches hostels were helping eighteen thousand men each month.[52]

Chaplains often found themselves acting as advocates of soldiers within the Army system, and this eventually resulted in the creation of a 'Morale Section' within the Chaplains Department.[53] Closely related to this was the work of home-based chaplains like Daniel Wakeley (Churches of Christ), who were part of specialised teams to help educate recruits about the dangers of venereal disease, which continued to be an enormous problem to the armed services. Wakeley reported to his Chaplain General of having given one hundred and fifty-one well-received lectures in one hundred and fifty-eight days and distributing hundreds of pounds' worth of literature.[54] The campaign, however, did not always run smoothly. The Chaplains General were concerned that the anti-VD pamphlets might be distributed to the Australian Women's Army Service, no doubt feeling that young servicewomen should not be exposed to such unsavoury details.[55] Their concerns, though well-intentioned (some would say patronising), failed to acknowledge that for every case of male infection a female was also involved, and that the young women might not be as innocent as they thought.

Moral issues and evangelistic priorities caused the state and federal boards of the United Churches in 1943 and 1944 to produce a number of pamphlets for their chaplains to distribute among the troops. In

---

52   Ken Manley, 'Carry On! Victorian Baptists and World War Two,' *Our Yesterdays*, 16 (2008), 37.
53   'Military History of the Australian Army Chaplains, Para 5, undated MS typescript, AWM 54, 177/2/1/1.
54   Gladwin, *Captains of the Soul*, 144.
55   Chaplains General' Conference, *Minutes*, May 1944, March 1945, in Tippett, 'Australian Army Chaplains,' 136.

April 1943 Brooke reported that he had obtained ten thousand copies of a pamphlet entitled 'Towards the Mark' for distribution to servicemen and women. Similarly, Albiston announced he had procured five thousand copies of the pamphlet 'Sex – Some Facts and Problems' for use by Air Force chaplains. They also agreed to make available evangelistic pamphlets written by Baptist chaplain John Ridley, entitled 'Hero among Heroes' and 'Over the Top'.[56] One month later the Victorian State Board asked the FUCCB to produce a pamphlet on 'Christian Verities,' including brief and clear statements concerning God, Man, the Scriptures, Immortality of the Soul, the Value of Prayer, the Brotherhood of Man and the Church.[57] This was followed in 1944 by an 'urgent' request for a pamphlet on the 'Foundations of our Faith'.[58]

In addition to such specially commissioned literature chaplains benefited greatly from the support they received from the British and Foreign Bible Society, which provided Bibles and New Testaments free of charge for distribution to servicemen and women. Following advice from chaplains that many servicemen did not have even a rudimentary understanding of the Bible, the Society in 1941 produced an introduction to the New Testament which was also made available for free distribution by chaplains.[59]

Chaplains also became involved in educational activities which eventually led to organising 'Request Hours' where they would address issues that concerned general morale and wellbeing. These were similar to the British Army's 'Padre's Hour', and soon after that term began to be used by Australian chaplains. After the war they became known as Commanding Officer's Hours, or CO's Hours, and were delegated to chaplains, who were mostly free to choose topics related to morals, ethics and religion.[60]

Concerns about issues confronting servicewomen led to 'a special relationship between the Army Chaplains Department and the women's services'. Several conferences were held to determine chaplaincy

---

56 VSUCCB, *Minutes*, 5 April 1943.
57 VSUCCB, *Minutes*, 24 May 1943.
58 VSUCCB, *Minutes*, 24 April 1944.
59 *Argus*, 29 April 1940, 7.
60 'AAChD – History 1939–46, United Churches,' in 'Military History of the Australian Army Chaplains,' undated MS typescript, AWM 177/2/1/1, Para 5.

policy with regard to servicewomen.⁶¹ In 1943 the Chaplains General sought advice from the Australian Women's Army Service concerning a proposal that women should be appointed as assistants to chaplains. A report by Captain Alma Hartshorne indicated that this was not necessary and that members of the Army Women's Service were happy with the current arrangements. Nevertheless, on 22 November, Major Kathleen Deasey was appointed liaison officer to coordinate chaplains' visits to AWAS units and members in civilian prisons.⁶²

Home-based chaplains also ministered in the internment camps that held German, Italian and Japanese prisoners of war, as well as civilians from those countries. Three were located near Hay, NSW, and accommodated refugees and internees, most of whom were Jewish. They are probably best known for being home to the famous 'Dunera Boys' who came to Australia in 1940 aboard SS *Dunera*. There were around one thousand German and Italian civilians in each of the camps. Most were intellectuals who were perceived as security risks. Despite their treatment being in accordance with the Geneva Conventions, their physical and emotional isolation from the outside world and their families was traumatic and lacking in compassion.

Nevertheless, those camps were well remembered by former internees for the ministry of Chaplain Ivan Alcorn (Churches of Christ), who ministered to all regardless of religion. Two of them corresponded with Tippett in 1987⁶³ and spoke of how Alcorn conducted regular ecumenical services, provided newspapers and other items that made life more bearable, and whose presence reminded them of another kindlier world: 'Most of us regarded his visits more for their value as a sign of life from the outside world, rather than a promise of everlasting life', one of them reported. The other indicated that Alcorn was responsible for the conversion of significant numbers of internees to active Christian faith.⁶⁴

In the military training camps, however, lack of interest in church services proved to be a perennial problem and source of discouragement

---

61 'AAChD Between the Wars," in 'Military History of the Australian Army Chaplains,' undated MS typescript, Australian War Memorial, 54, 177/2/1/1, Para 10.
62 Report on Visits to AWS Units, AWM 54, 177/2/1/1.
63 These letters, with others used in his thesis, were lost following Tippett's death in 2012.
64 E. Seaton and G. Leister as cited in Tippett, *Australian Army Chaplains*, 158.

for chaplains. Yet, like their World War I predecessors, they found that things changed when they entered the combat zones. Alan Prior and Frank Starr (Baptists) both commented that having found little interest in their services in the camps at home, 'the nearer the front and the fighting, the greater the attendance at services and the more requests [there were] to conduct services of worship'.[65] In this respect the trans-denominational character of World War I chaplaincy, much favoured by UC chaplains, emerged again.

The War in the Pacific

The Southwest Pacific Campaign for Australia began on 8 December 1942, when Prime Minister John Curtin announced the declaration of war against Japan.[66] It heralded the start of the most dangerous period in Australia's history and led to an unprecedented build-up of military forces, including UC chaplains. Moreover, it also introduced a new dimension of tactical capability as the Army developed tactics related to jungle environments rather than open battlefields. Gladwin notes that 'the nature of much of the war in the Pacific campaign revolved around small groups of men deployed in mobile tropical jungle warfare'.[67] This meant that life for most of the UC chaplains in this theatre of war was spent ministering to small groups, with relatively little contact with other chaplains. It was a very different environment from that of World War I and the Middle Eastern Campaign of 1940–1942, where vast numbers of troops were massed in relatively small areas and UB/UC chaplains worked in close contact with those from other denominations.

This new operational environment, more than anywhere, demanded an ecumenical rather than denominational focus in worship services. Chaplains discovered that the most effective services were those held before or after a battle. Harold Norris (Churches of Christ) reported that 'few soldiers absented themselves from services after the battalion had been in action'.[68] However, both he and Eric Hollard (Churches

---

65   Alan Prior and Frank Starr, as cited in Tippett, *Australian Army Chaplains,* 131.
66   John Curtin, Declaration of War on Japan, NAA C102, 1552510.
67   Gladwin, *Captains of the Soul,* 121.
68   Harold Norris, as cited in Tippett, *Australian Army Chaplains,*' 163.

of Christ) objected to compulsory church services, considering them to be 'religion by force'.[69] Other chaplains, like Frank Starr, found that smaller, personal services were particularly effective. Starr recalled how he had conducted a service on each of the artillery guns in his area, and on Easter Sunday 1943 held sixteen services, including communion, with groups of soldiers where they sat.[70]

Conducting religious services, though clearly the most conspicuous duty of chaplains, was not their most pressing and appreciated activity. Individual pastoral care and counselling was, and still is, a chaplain's most consuming role. But it also took its toll on the chaplains. Malcom McCullough (Baptist),[71] in a letter to his wife from New Guinea, confided:

> Sometimes I feel as though I am unable to cope with all that has to be done ... I just can't do it ... my mail piles up ... awfully tired and sleepy ... I fought hard for the lad, but did not succeed ... It quite depressed me.

Still, McCullough was able to report that 'he had not departed from the essential and fundamental concepts of the gospel', even though, as he reported later, 'he felt starved' because he had not been receiving copies of the *Baptist* and the *Christian Evangelist* which had previously enriched his spiritual life. He, like most UC chaplains, whose theological orientation tended to be at the conservative evangelical end of the spectrum, maintained his own personal spiritual vitality through regular prayer and Bible study.[72]

Both Brooke and Albiston raised concerns about the age and fitness of chaplains. They worried about the resilience of older men working under relentless pressure in wartime conditions and stressed the need for younger men to replace them.[73] This was easier for Brooke, who reported on several occasions that the number of UC chaplains in the Army exceeded the allocation based on census figures. In September

---

69 Eric Hollard, as cited in Tippett, *Australian Army Chaplains*, 163-164.
70 Frank Starr, as cited in Tippett, *Australian Army Chaplains*,' 163.
71 Later to become United Churches Chaplain General.
72 Malcom McCullough, *Letters to his wife*, in Tippett, 'Australian Army Chaplains,' 173.
73 VSUCCB, *Minutes*, 5 April 1943, 3 July 1944, 12 February 1945.

1943, for example, he reported that he 'was 16 chaplains over strength',[74] whereas Albiston, in May of that year, had indicated the 'position in RAAF was the reverse and he needed more chaplains'.[75] However, less than a year later the position had changed significantly and Albiston stated that he would now only accept applications from clergy aged between 28 and 40 years of age.[76]

As in World War I, chaplains in this theatre were heavily involved in ministry to the dead and dying. Many chaplains volunteered to act as stretcher bearers, including Doug Christian (Salvation Army), who was described in his battalion's official history as 'Christian in name and nature'.[77] Robert Helmore, following 'four days of the toughest, grimmest fight the battalion had experienced' described how he helped identify and bury the dead, then wrote personal letters to their families. He reported that most of the religious services he conducted were at gravesides, but he remembered particularly one quiet, reverent service in a tent where he and a group of soldiers 'gathered at the table of remembrance, to break bread and drink the cup of the new Covenant'. The only light they had issued from two hurricane lamps and the bright moonlight, but 'if ever the presence of Christ was real to men, we knew it at this service', he said.[78]

In addition to conducting funeral services, chaplains were required to mark out and record cemeteries in the jungle, as well as collecting the personal effects of the deceased and sending them to the appropriate authorities. Norris reported that between November 1942 and March 1943, he had buried eleven officers and one hundred and fifty-one other ranks, as well as ministering to thirteen officers and two hundred and seventy other ranks who were wounded. He also mentioned that in accordance with the tradition of practical ecumenism 'No notice was taken of different denominational customs'.[79] In this respect it is important to note that, contrary to civilian practice at that time,

---

74 VSUCCB *Minutes*, 2 September 1943.
75 VSUCCB *Minutes*, 24 May 1943.
76 VSUCCB *Minutes*, 24 April 1944.
77 2/4th Australian Infantry Battalion Association, *White over Green* (Sydney: Angus & Robertson, 1963). 231.
78 *Victorian Baptist Witness*, 5 December 1942, 4.
79 Harold Norris, as cited in Tippett, 'Australian Army Chaplains,' 184.

war cemeteries made no distinction between religious affiliation or lack thereof.

This is further evidence of the trans-denominational ethos of Australian military chaplaincy on operations, and a mutual acceptance that was far ahead of that which characterised civilian clergy. Despite the obvious theological and ecclesiastical differences in such a diverse group, what records remain constantly highlight the comradeship that existed among the chaplains and their respect for each other as servants of God. In the crucible of war, issues of theology and churchmanship became unimportant compared with the overwhelming need to minister to troops living in a man-made hell. UC chaplains, who, in their home environments may have felt they had little in common with clergy from the major churches, soon found the opposite to be true. The experience of war taught them that the presence of Christ is revealed not in chaplains' denominational accreditation but in the depth of their faith, hope and love. Brooke, who presided over the most disparate group of chaplains, spoke of this in his Post War Report, recording his deep appreciation of the unity and loyalty shown by all his United Churches chaplains.[80]

Regardless of the isolation experienced by most chaplains due to the nature of this theatre of war, Chaplains Department policy again was to post one Anglican, one Roman Catholic and one Protestant chaplain to each Brigade, even though its units and sub-units might be scattered over a large area. However, in 1943 the three Protestant Chaplains General agreed that 'for administrative purposes the work of chaplains of the Presbyterian, Methodist and United Churches [could be performed by] one or other acting for the three groups'.[81] This was a momentous decision, which formally sanctioned the practical unity already evident on the ground and prefigured what was later to become standard policy in all three services.

Nevertheless, chaplains' feelings of isolation were compounded by Senior Chaplains being located at Divisional Headquarters that were often too far away for them to be of much help. There were exceptions. Frank Starr made special reference to the time that his Senior Chaplain,

---

80 Allen Brooke, in *AAChD Post War Report* (Canberra: Army Office, 1945).
81 'AAChD – History 1939–46, United Churches,' in 'Military History of the Australian Army Chaplains,' undated MS typescript, AWM, 177/2/1/1.

James Methven (Churches of Christ), spent with him in Aitape-Wewak, describing him as 'a tremendous blessing'.[82] Methven was later Mentioned in Despatches and awarded the MBE:[83] 'For courageous devotion to duty ... Continuously in the forward areas ... an inspiring example of devotion to duty ... [having earned] the respect and affection of thousands of men in the Division'.[84]

Hugh Ballard (Congregational) was awarded the Military Cross for his courage and inspiring leadership during an attack by 2/11 Australian Commando Squadron in Labuan on 16 June 1945. While under fire, and though a non-combatant, he personally organised the evacuation of a number of wounded soldiers and assisted in the destruction of the enemy position that caused the casualties by pointing out its location to assault troops. His citation made special mention of his coolness, courage and disregard for personal danger, which 'greatly inspired the men who were assisting him'.[85]

In addition to Methven, four UC chaplains were Mentioned in Despatches: Robert Pickup,[86] Thomas Keyte[87] and Arthur (Harry) Orr[88] (Baptists), and Frank Fewster[89] (Churches of Christ). Orr's official report on a chaplaincy visit he made to the scattered detachments along the Morobe–Amboga Signal Line, while posted to Lae Base Sub Area Headquarters, gives a helpful insight into the way that chaplains operated in this unique theatre of war. Orr stated that his objective was 'To minister to all Protestant men in spiritual matters ... to take comforts and literature to all' regardless of denomination, and 'To have a few hours of music, discussion, games and fellowship at each station'.[90]

He set out on his journey on 19 May 1945 taking with him, in addition to his personal gear, 'about 100 lbs of the very best literature available, 100 lbs of comforts including chocolates, soaps, shaving

---

82 Frank Starr, as cited in Tippett, 'Australian Army Chaplains,' 190-191.
83 'Military History of the Australian Army Chaplains,' United Churches, AWM 54, 177/2/1/1.
84 Methven J.O. Military Record, NAA B883, QX24324, 7, 9, 23.
85 Ballard H.R. Military Record, NAA B883, SX22714, 31.
86 Pickup R.S. Military Record, NAA B883, NX156703.
87 Keyte, T.F. Military Record, NAA B883, VX64453.
88 Orr A.H. Military Record, NAA B883, NX200382.
89 Fewster F.M. Military Record, NAA B883, WX37414.
90 Harry Orr, *Report on Visitation of Signal Line Morobe – Amboga Area* (Baptist Union of Victoria Archives, Melbourne), 1.

creams, toothpastes, games, sox, [*sic*] toothbrushes as well as many other articles, obtained from A.C.F. [Australian Comforts Fund], Y.M.C.A and Salvation Army sources'. He also included a supply of New Testaments for distribution with the books. Furthermore, in the best traditions of holistic chaplaincy care, he took his piano accordion and a wireless set to provide entertainment.

He described the journey as 'arduous in many places' and after rain, potentially 'impassable ... and difficult to get through without danger'. He completed his journey on 7 June at Dobadura, where he mingled with the personnel and conducted a 'general service' at Higoatura, before being air-lifted back to his base at Lae. The trek had taken him nineteen days. He described his *modus operandi* thus:

> My general plan was to visit one sig.[nal] station a day, and remain overnight. I endeavoured to arrive at each station about 1600hrs., distribute the comforts while it was still light and generally settle in. After tea we would settle down to an evening of music, yarning, and, if so desired, games etc. Usually we talked into the small hours. In the morning, after breakfast, we would ... sing some of the old time hymns, accompanied by the piano accordion, offer a prayer, read a selected passage from the Bible and then, instead of an address, introduce some aspect of the Christian faith for discussion.[91]

His report also included his assessment of the men's morale, something commanders tend to value highly from their chaplains, including suggestions about ways in which this might be improved. He reported that he 'found the men most ready to speak about spiritual things and desirous of Christian Literature and Testaments ... The dominant issue in all their discussions about religious matters [he said] was the search for a rational background to the faith'. He concluded his report by saying he felt this had been accomplished.[92]

---

91 Orr, *Report on Visitation of Signal Line*, 2.
92 Orr, *Report on Visitation of Signal Line*, 4.

Prisoners of War

Between 1942 and 1945 thirty-four Australian Army and one Australian Navy chaplains were among more than twenty-two thousand Australians captured by the Japanese and held in prisoner of war camps scattered between Burma and Japan.[93] None of them were UC chaplains,[94] except perhaps for Alan Garland (Churches of Christ) who, like many of his World War I forebears, chose to enlist in the ranks of the Australian Army Medical Corps, rather than in the Chaplains Department. Garland, who was initially incarcerated in Changi camp and later died in Borneo on the infamous Sandakan Death March,[95] served unofficially as a chaplain to his fellow POWs but without formal rank and status.[96] The only official UC Chaplain on full-time duty to die during World War II was Horace Willings[97] (Congregational), who died in Australia of illness on 20 August 1942.

Navy and Air Force Chaplains in World War II

As in World War I, no United Churches chaplains served with the Royal Australian Navy during World War II. The Navy Board continued its policy of appointing chaplains relative to the reported denominational affiliations of serving members of the Navy, arguing that the small number of adherents of United Churches were catered for by existing Protestant chaplains.

Nevertheless, the Victorian State Board did make some attempts to challenge this. In November 1943 it agreed that Brooke and Albiston should meet with the Minister for Navy to urge him 'to consider the

---

93 Gladwin, *Captains of the Soul*, 129; Rowan Strong, *Chaplains in the Royal Australian Navy, 1912 to the Vietnam War* (Sydney: UNSW Press, 2012), 196; Cameron Forbes, *Hellfire, The Story of Australia, Japan and the Prisoners of War* (Sydney: Macmillan, 2005), 502-503.
94 Most of the Australians captured by the Japanese were from the Australian 8th Division, which was raised before the major expansion of the Australian Army and RAAF beginning in 1943, when the majority of United Churches chaplains were appointed.
95 Paul Ham, *Sandakan, The Untold Story Of The Sandakan Death Marches*, (Sydney: William Heinemann, 2012), 373.
96 Garland A.W. Military Record, NAA B883, VX32307, 6, 8.
97 Willings H.J.W. Military Record, NAA B884, N429336.

position of United Churches chaplains in the Navy'.[98] This proved to be a futile endeavour. Their reports to the Board meetings in February and April 1944 noted that it had not been possible to arrange such a meeting.[99] One year later, in July 1945, they reported that the Minister had agreed to meet with representatives of the FUCCB 'when in Melbourne'.[100] However, by this time they were sufficiently doubtful of the Minister's interest that they agreed 'not to wait but to pursue the matter with the Navy as urgent'. Even so, the September meeting simply recorded: 'No reply received'.[101]

The Air Force, however, was a different matter. From the original six chaplains serving when war was declared, the total complement reached three hundred and seventy-two by the war's end. The huge expansion of the RAAF, following the entry of Japan into the war, opened the door for the United Board/United Churches to nominate 29 new chaplains.

As previously mentioned, the appointment of four Staff Chaplains in August 1940, one from each of the major churches, began the process of formalizing the RAAF Chaplain Branch. The Chief of the Air Staff, Air Chief Marshall Sir Charles Burnett, took a keen interest in the establishment of the branch and supported this move. Initially, the Staff Chaplains were given the rank of Squadron Leader and though having some general administrative authority, their authority only applied to those of their own denominations. This was essentially a repeat of what had happened a generation earlier in the creation of the Army Chaplains Department. It also emphasised the principle that chaplains belong primarily to their Church that nominated them, and they remain under its discipline in addition to being subject to military authority.

As previously noted, the first four Staff Chaplains appointed were Joseph Booth (Anglican), Kenneth Morrison (Roman Catholic), Arthur Davidson (Presbyterian) and Thomas Rentoul (Methodist).[102] Booth suggested that a Staff Chaplain for the United Board should also be appointed. However, with echoes of 1915 and the proposal for an

---

98   VSUCCB *Minutes*, 16 November 1943.
99   VSUCCB *Minutes*, 21 February 1944; and 24 April 1944.
100 VSUCCB *Minutes*, 30 July 1945.
101 VSUCCB *Minutes*, 30 July 1945; and 24 September 1945.
102 Application for Appointment as Staff Chaplain to RAAF – Rev T.C. Rentoul and A.I. Davidson and Rev Fr Morrison and Rt Rev J.J. Brook, NAA A705, 36/1/109.

OPD Chaplain General for the Army, the main opposition came from the Presbyterian and Methodist Staff Chaplains,[103] who argued that the low numbers of airmen claiming affiliation to United Board churches did not warrant such an appointment. Consequently, Booth withdrew his suggestion.[104] However, forty new chaplains were recruited in late 1940, including Walter Albiston (Congregational) who in 1942 was advanced to Staff Chaplain (United Churches).[105]

By then the number of UC chaplains in the Air Force had risen to ten, and by the end of the war had reached twenty-nine.[106] One of them, Douglas Riley (Congregational), was appointed Area Command Chaplain in Morotai in 1945.[107] Their role was very similar to that of their Army counterparts: conducting worship services, religious instruction, providing pastoral care, visiting the sick and wounded, burying the dead, writing to next-of-kin, censoring mail, and distributing 'comforts'. Those posted to operational stations in war zones were subject to similar dangers as their Army brethren. Not having had previous war experience, the RAAF's Chaplaincy Branch went to war with no established doctrine of active-service chaplaincy except what had been gleaned from those, like Staff Chaplains Booth, Davidson and Rentoul, who had served as Army chaplains during World War I. But, like that previous generation, they learned as they went.

Many found it a lonely experience and complained of the lack of support from their churches, such as James Sweet (Presbyterian), who confided, 'For eighteen months in the north I was a very lonely Padre – forgotten by my church'.[108] Their situation, though often isolated, was probably less so than for many Army chaplains in that theatre of war. They were posted to operational squadrons located at major bases, rather than scattered units and sub-units spread over areas of impenetrable jungle, and usually had easier access to other chaplains.

103 See p.49. Despite sharing the same name and having similar opinions about OPD/UC participation, WWI Presbyterian Chaplain General John Lawrence Rentoul and WWII Methodist Staff Chaplain Thomas Rentoul were not related.
104 Davidson, *Sky Pilot*, 2-4.
105 Air Force Regulations to Apply for Appointment of 5 Staff Chaplains, NAA A14487, 46/AB/6257.
106 Davidson, *Sky Pilot*, Statistical Summary RAAF Chaplains – WW II.
107 Congregational Union of NSW Yearbook 1945, Uniting Church Archives, Sydney.
108 J.R. Sweet, cited in Davidson, *Sky Pilot*, 7-1.

Davidson describes a typical day in the life of an operational squadron chaplain. After his personal morning devotions, he would spend time censoring mail, counselling airmen, visiting the wards at the nearby medical unit, distributing 'comforts', visiting other units nearby, and writing to the families of the men he served. In the evening, he would usually participate in a recreational activity, such as leading a debate or giving a talk on some relevant subject. Some evenings he might conduct a church service in addition to those he led on Sundays.[109]

Arthur Wilkins (Baptist) found the RAAF chaplain's role to be quite complex and included things not usually associated with being a minister of religion. He described his developing experience thus:

> I learned that a chaplain's position had many aspects. He was a counsellor and adviser to personnel and their families, a human problem solver, a mediator, an educational resource, a librarian. He was given administration responsibilities such as censoring letters or control of squadron funds. Anything, in fact that did not fit nicely into the RAAF scheme of things was a job for the Padre.[110]

From little more than a handful of chaplains on the outbreak of World War II, the RAAF Chaplain Branch grew exponentially to meet the demands of a rapidly expanding service. Wherever airmen and women served, their chaplains served and supported them. But it was the entry of Japan into the war in December 1941 that opened the door for the United Board/United Churches to make their contribution to this growing force. As with the Army, the major churches found it increasingly difficult to meet their quotas of chaplains, and UC chaplains filled the gaps, almost doubling the number required under the policy of proportionality and establishing themselves henceforth as a permanent presence in that service.

---

109   Davidson. *Sky Pilot,* 7-2.
110   Arthur Wilkins, *Life as I see It* (Brighton: Association for the Blind, 1988), 83-84.

Onward and Upward

Whereas World War I became the defining moment of Australia's nationhood, World War II was its greatest challenge. The entry of Japan into the war brought the conflict to the Australian homeland and threatened its very survival. Bob Wurth presents a vivid image of the desperation of those times when he refers to Prime Minister Curtin, at the height of the crisis, being 'stranded by floodwaters in the barren wilderness of the Nullarbor, connected to the world only by the dots and dashes of morse code, while the Australian territory of Rabaul ... fell to a massive Japanese invasion'.[111] General MacArthur's assessment of Australia's defences, given soon after, described them as 'weak to an extreme ... the bulk of its ground forces in the Middle East ... its air force equipped with almost obsolete planes ... its navy [with] no carriers or battleships'.[112]

The campaign in the Southwest Pacific thus became a fight for national survival unmatched by previous wars and campaigns. It led to the greatest mobilisation of personnel and resources – material, economic and spiritual – in the nation's history. Australia's military chaplains were a vital part of the fight, and those belonging to the United Board/United Churches contributed considerably more than the relatively insignificant size of their denominations required.

The response to the declaration of war by the churches matched that of the nation, well expressed in Prime Minister Menzies' description of it as a 'melancholy duty'. There was little evidence of the patriotic fervour and unbounded enthusiasm for battle that greeted Australia's entry into World War I. The response in September 1939 was that of grim acceptance of a reality that people previously believed would never happen again, and a determination to do whatever had to be done to end it. Church leaders displayed little if any of the 'God of battles' rhetoric that so characterised the pronouncements of August 1914. Even so, Pastor Clemens Hoopman, President of the Evangelical Lutheran Synod of Australia, still spoke of war as 'a terrible scourge for the nations of the world ... sent by God for the punishment of the ungodly and the

---

111   Bob Wurth, *The Battle for Australia*, (Sydney: Pan Macmillan, 2013), xi.
112   Douglas MacArthur, cited in Wurth, *The Battle for Australia*, 264.

chastisement of his people'.[113] It is doubtful, though, that such simplistic moral judgements were shared by chaplains faced with the horrific reality of battle.

As in World War I, the churches ensured that their chaplains were present wherever the nation's servicemen and women were deployed: in training camps at home, forward mounting bases, operational areas and on the high seas. Once again, the United Board/United Churches rose to the challenge, offering the services of those clergy who already belonged to the Citizens Forces and others who were ready to volunteer. For the first two-and-a-half years they were relatively few as chaplaincy numbers were kept strictly in accordance with the principle of proportionality. But the threat from Japan and the opening of the Southwest Pacific campaign led to such an expansion of the Army and Air Force that the major churches found it increasingly difficult to meet their requirements, and looked to the United Churches to make up the deficit.

With the exception of the Salvation Army, clergy of its member denominations were much freer to volunteer than those from the more hierarchical denominations whose bishops, already under pressure from the war effort and desperate to maintain their own work at home, were sometimes reluctant to release them, adding to their denominations' recruiting difficulties. This helps explain why the United Churches provided almost four times the number of chaplains that strict adherence to census figures required.

The five themes that emerged through the experiences of World War I chaplains: incarnational ministry, ardent evangelicalism, practical ecumenism, servant leadership and continuity of service, were again present in World War II, as this chapter has demonstrated. Their chaplaincy praxis differed little from that pioneered by their World War I predecessors, except for the introduction of 'Request Hours' which, as CO's Hours, became a regular feature of chaplaincy after the war. Concerns about excessive drinking and sexual morality figured strongly in chaplains' concerns: sometimes even in relation to their fellow chaplains, as recorded by Eric Hollard (Churches of Christ) who deplored

---

113 Clemens Hoopman, in *On Service with the Men and Women of the Evangelical Lutheran Church* (Adelaide: the Service Commission of the Evangelical Lutheran Church, Undated), 3.

the behaviour of some brother chaplains, in particular 'their use of the bar in the Officers' Mess'.[114]

The most noticeable omission in the records left by World War II chaplains is any serious reflection on issues relating to the theology and morality of the war. World War II was undoubtedly the most horrific and destructive war in the history of humankind, and most of its victims were civilians. An estimated total of around fifty million people perished, of whom most were civilians. The principles of the Just War Theory were violated not only by the Axis powers but also by the Western Allies, most obviously in the deliberate targeting of civilians in mass bombing raids, culminating in the atomic bombs dropped on Japan.

Gladwin, however, observes that war, for most chaplains, 'Served to strengthen their convictions about the depravity of human nature and the necessary power of the Christian gospel to change lives'.[115] Even so, for McCullough (Baptist), observing the fortitude of Australian soldiers, it caused him to 'feel a bit at sea in saying that man is so totally depraved'.[116] The extent of human suffering caused many to question the idea of a God who is omnipotent and whose essential being is love. Nevertheless, it appears that most chaplains, like their churches, accepted the reality of having to live in a fallen world where the only choice was between something bad, such as the bombing of civilians, or something worse like the triumph of Nazism and Japanese militarism. It came down to a choice between the lesser of two evils. Gladwin perceptively observes: 'The visceral impact of the war and its legacy of suffering meant that few chaplains had the luxury of armchair theologising. More nuanced theological reflection would have to wait'.[117]

In this sense the chaplains were somewhat different from Michael McKernan's assessment of Australian clergy in World War I, who 'Never broke themselves clear of the events ... ever reacting rather than acting'. McKernan argues that at a time when Australians needed 'calm words of caution to help them subdue their ... romantic view of war;

---

114 Eric Hollard, as cited in Tippett, 'Australian Army Chaplains,' 194.
115 Gladwin, *Captains of the Soul*, 155.
116 Gladwin, *Captains of the Soul*, 156.
117 Gladwin, *Captains of the Soul*, 158

clergymen added to the rhetoric of hysteria'.[118] For the chaplains who had witnessed its horror, the aftermath of World War II was not accompanied by visions of a now enlightened world that would ensure that such tragedy would never happen again, but rather a grim acceptance of the reality of fallen human nature. To some extent this was offset by optimism that the creation of the United Nations in 1945, whose aim was to prevent future world wars, would succeed where the previous League of Nations had failed. Nevertheless, the emergence of the Soviet Union as a new superpower, implacably opposed to the western democracies, did little to allay fears that the world might again be plunged into another holocaust, as was articulated by Winston Churchill in his famous reference to 'an iron curtain' having descended across Europe.[119]

What records exist of UC chaplains' reflections and experiences during World War II, as well as references to the words and writings of their denominational leaders, indicate that their overriding concern was essentially pragmatic: humankind was confronted with an unspeakable evil in German and Japanese totalitarianism, which had to be defeated whatever the cost. Their role was to provide whatever pastoral and spiritual support they could to those who did the fighting and, like their predecessors of World War I, that is what they did.

Yet, despite its unspeakable evil, the war did produce something they treasured for the rest of their lives: the memory of ordinary men and women who, when faced with the worst that humankind can bring forth, responded with a level of endurance and self-sacrifice that transcended its horror. Harold Law-Davis (Baptist) wrote:

> A common danger has made us realise, as nothing else could ... [what] we have gained in the experience of comradeship. It goes deeper and affects more profoundly our lives than any other thought or attachment ... We differed in every way, mentally, physically, spiritually, politically. But we had one thing in common, a mutual respect for and trust in each other, the fruit of a new comradeship. It was something the war had given us.

---

118 Michael McKernan, *Australian Churches at War: Attitudes and Activities of the Major Churches 1914–1918* (Sydney and Canberra: Catholic Theological Faculty and Australian War Memorial), 172.
119 This speech was delivered in March 1946 at Westminster College in Fulton, USA.

It is something we must preserve when the war has gone. And strange, but true, at the basis of the Christian faith is comradeship – God, the Comrade of man. "The Word was made flesh and dwelt among us."[120]

---

120   Harold Law-Davis, 'Padres Ponderings'. *Light Diet* (January 1945), 17.

CHAPTER SEVEN

The Cold War:
The Stone Stands Fast

WORLD WAR II, WHICH BEGAN without the boundless enthusiasm that marked the start of World War I, ended without its optimistic hope of a world in which war would be no more. The post-war generation grew up under the shadow of a third world war that would likely bring an end to human civilization as we know it. Josef Stalin, in a speech prior to the elections to the Supreme Soviet in February 1946, 'claimed that capitalism made war inevitable'.[1] History named it the Cold War: 'An intense economic and political rivalry just short of military conflict, involving the Soviet Bloc on the one hand and the Western Powers on the other and lasting from 1945 to the collapse of the Soviet Union in 1990'.[2] Peter Firkins expressed it well:

> Those who survived the First World War, and in their innocence had hoped for an era of peace and plenty, had been granted neither. Those who were discharged in 1945, and their sons after them, have lived through a phenomenal era of a constantly developing world economy but a constantly threatened world security.[3]

---

1 Jeremy Isaacs & Taylor Downing, *Cold War* (London: Bantam, 1998), 29.
2 'Cold War,' *Macquarie Dictionary* 2005, 4th edn., Macquarie University, Sydney, 374.
3 Peter Firkins, *The Australians in Nine Wars, From Waikato to Long Tan* (Sydney: Pan, 1982), 443.

The Cold War shaped Australia's defence policy for the next three decades, especially after 1950 when the prospect of communism dominating post-colonial Asia loomed large in global strategic thinking. In 1946 the first post-war review of Australia's strategic circumstances led to a policy of forward defence that would see Australian forces over the next quarter century fighting in Malaya, Borneo, and Vietnam. While giving priority to the need for post-war reconstruction, the government endorsed the need for a strong defence capability, which for the Army meant the raising of a regular force[4] whose size would be similar to that of the CMF.[5] For Australia it was a time to stand fast against a new international threat. For the United Churches it was a challenge to maintain its reputation as an integral part of Australian military chaplaincy.

Post-War Army Chaplaincy

The Army Chaplains Department, which during World War II 'had overcome past inertia to be a real element of the Army … [and] had also sponsored the most ecumenical outreach the nation had ever seen,'[6] returned to a peacetime establishment based again on the principle of proportionate representation, which the major churches found easier to achieve in a much smaller Army. Conscious of the two decades of 'haphazard' inefficiency that followed World War I, the Chaplains Department was determined to consolidate its position in the increasingly professional post-war army, and the United Churches, now with its own Chaplain General, was well placed to be part of it.

Even so, Douglas Abbott argues that the Chaplains General, Allen Brooke in particular, still saw themselves as the appropriate body to lead and administer the Chaplains Department. He claims that Brooke was threatened by the idea of 'a cadre of Regular Army chaplains with long service, who would be knowledgeable about the army system and gain

---

4   To be designated the Australian Regular Army.
5   Jeffrey Grey, *A Military History of Australia* (Melbourne: CUP, 2008), 198-199.
6   Michael Gladwin, *Captains of the Soul. A History of Australian Army Chaplains* (Sydney: Big Sky, 2013), 174.

acceptance within the military community'.⁷ Michael Gladwin agrees, indicating that the feeling among those who later became Regular Army chaplains was that their part-time Chaplains General were out of touch with the needs of a 'professionalising post-war Army'.⁸ It was an issue that continued to manifest itself and was only finally resolved in 1990.

The creation of a regular army in September 1947, supported by the reorganised CMF, changed the functional environment of the Australian Army, which since Federation had been a part-time force with only a small number of permanent members. It opened the way for an emerging class of professional leaders who would eventually dominate what had traditionally been a citizens army. Included among these were a number of chaplains serving full-time on short service commissions. A similar process was emerging in the Air Force,⁹ and Albiston reported that short service commissions of four years' duration were now a possibility for Air Force chaplains.¹⁰ Brooke indicated that the Army, though not prepared to create permanent positions, would extend periods of continuous full-time service, if the situation warranted it. This had happened with Harry Clark (Churches of Christ) who had then completed six years of full-time service.¹¹ It opened the way for the gradual emergence of a professional chaplaincy group and long-term career chaplaincies.

Despite Abbott's criticisms, one Chaplain General at least was prepared to challenge the *status quo*. In 1945 Alexander Stevenson (Presbyterian) recommended that there should be only one Protestant Chaplain General rather than the current three. It was a logical proposal because it would have given the Protestants more leverage against the influence of the Anglican and Roman Catholic Chaplains General, who were sometimes accused of adopting a 'divide and conquer' strategy towards the Protestants.¹² It was also a generous proposal in that

---

7   Douglas Abbott, 'In This Sign Conquer: The Chaplains General of the Australian Army 1913–1981,' unpublished manuscript, 414.
8   Gladwin, *Captains of the Soul*, 255
9   Navy chaplaincy had always been largely full-time.
10 VSUCCB *Minutes*, 12 October 1948.
11 VSUCCB *Minutes*, 15 March 1949.
12 This was a frequent complaint made by Protestant chaplaincy leaders during the writer's term on the Religious Advisory Committee to the Services.

it may have resulted in Stevenson himself losing his appointment as a Chaplain General to one of the other two Protestants.

However, it was deemed unacceptable to the churches concerned and likely to raise numerous difficulties.[13] What those difficulties were was not specified, but it is possible that the other two Protestant Chaplains General influenced this decision: it would not have been easy to give up the status and pecuniary benefits of being a Chaplain General. But it should also be remembered that the United Churches, after thirty years on the margins of military chaplaincy, had only recently achieved their desired goal of equal standing with the others. Such a demonstration of ecumenical cooperation, no matter how logical and potentially empowering for the Protestants overall, was evidently ahead of its time.[14]

Ten years later, another proposal to increase efficiency and ecumenical cooperation emerged when it was agreed that one of the Chaplains General should be appointed as full-time secretary of the Conference. The Adjutant General agreed to grant an additional two days per week and Brooke, as current secretary, in October 1956 took up this expanded role, doubling his existing allocation of eight days per month.[15] He served effectively in this position until his retirement in 1964, enjoying the full confidence of his fellow Chaplains General, none of whom ever proposed there should be a change while he was available. Indeed, Bishop John Morgan (Roman Catholic), reminiscing on his experience as Roman Catholic Chaplain General, made particular reference to Brooke's significant contribution in his secretarial role.[16] It was a feeling that was obviously shared by the other members of the Conference, and was further evidence of the change in status of the once marginalised group that Brooke represented.

---

13 Abbott, 'In This Sign Conquer,' 68.
14 A similar proposal was put forward sixty years later by Rev David Griffiths (Baptist). It met the same end.
15 Abbott, *In This Sign Conquer*, 70.
16 J.A. Morgan, 'Interview with Bishop Morgan,' interviewed by Graham Downie, National Library of Australia, 24 July 1995. Audio, tape 3; Gladwin, *Captains of the Soul*, 165.

Post-War Air Force Chaplaincy

By 1945 Australia's Air Force had become the fourth largest in the world. Demobilisation, however, saw it reduced from a peak of one hundred and sixty-four thousand personnel in 1944 to eight thousand in 1948. Its chaplaincy branch also diminished significantly, from around two hundred and fifty to eight full-time and eighteen part-time chaplains. Of these, half were from United Churches: one full-time and eight part-time.[17]

Walter Albiston during his twenty years of service, which included six as Principal Air Chaplain, played a significant part in the transition of the chaplaincy branch from its wartime footing to a peace-time establishment.[18] His impact, like that of Brooke in the Army, gave United Churches chaplaincy a credibility at the higher levels of Air Force chaplaincy that ensured continuing equal standing with the major churches. Like Brooke, he was highly esteemed by his colleagues, who spoke of his 'diligence, impartiality and gracious ecumenical spirit, shown especially in his role as chairman of the Board of Chaplains'. The Board's minutes record:

> Whilst holding firmly to his own theological beliefs and viewpoints he could not tolerate bigotry, and by his brotherliness and friendliness did much to maintain the harmonious and efficient functioning of the Board of Chaplains.[19]

Consolidation and Growth

Throughout this period, the number of UC chaplains in both Army and Air Force fluctuated. Initially they continued to reflect the trend towards greater representation. In 1947 Brooke listed United Churches

---

17 Four Baptists, two Congregationalists, two Churches of Christ and one Salvation Army: see Davidson, Peter. *Sky Pilot, a History of Chaplaincy in the RAAF 1926–1990* (Canberra: Directorate of Departmental Publications, Department of Defence), Roll and Service Record RAAF Chaplains 1926–1990, unnumbered pages.
18 In 1953, the designation Staff Chaplain was changed to Principal Chaplain and the rank increased to Air Commodore; VSUCCB *Minutes.* 23 Mar 1953.
19 Royal Australian Air Force Board of Chaplains *Minutes of Meeting,* 4 February 1965; Davidson, *Sky Pilot,* 10-13.

as having four full-time and sixty part-time chaplains on the Army Active List.[20] In 1958 the number serving full-time increased to six, but part-time reduced to thirty-three.[21] Full-time representation then returned to four in 1969.[22] A variation in official policy occurred in 1957 when Brooke reported that the official establishment for United Churches had risen to 9 percent.[23] This may be partly explained by the inclusion of Lutherans in the combined United Churches census figures, but probably also reflects the fact that Brooke was now recognised as being responsible for the interests of all religious bodies not represented by the other four Chaplains General.[24]

Full-time representation in the Air Force also began to grow. In 1958 Albiston was invited to fill a vacant Presbyterian position, thereby doubling the United Churches allocation to two full-time chaplains. The Victorian State Board expressed its appreciation for what it described as a 'gracious gesture on the part of Principal Air Chaplain Russell',[25] although, in reality, it was part of an established trend whereby UC chaplains compensated for unfilled vacancies that might otherwise disappear from the chaplaincy establishment.

The effective administration of UC chaplains during this critical period was primarily due to Brooke and Albiston. Despite the creation in 1942 of the Federal United Churches Chaplaincy Board, in the postwar years the FUCCB had somehow ceased to operate. Consequently, on 25 November 1958 Brooke and Albiston met with representatives of the Baptist Union, Churches of Christ and Salvation Army, where Brooke reported that even though the Federal Board had not met 'for some years',[26] he and Albiston, along with other members, had continued to meet regularly with the Victorian State Board, thereby ensuring continuity of United Churches administration.[27] The result was an agreement to reconstitute the Federal United Churches Chaplaincy

---

20 Alan Brooke, Letter to the FUCCB, 23 July 1947.
21 VSUCCB *Minutes,* 27 May 1958.
22 FUCCB, *Minutes,* 28 June 1966; 12 August 1969.
23 VSUCCB, *Minutes,* 26 February 1957.
24 FUCCB, *Minutes,* 9 November 1965.
25 VSUCCB *Minutes,* 25 February 1958.
26 His report does not indicate when the Federal body stopped meeting.
27 FUCCB, *Minutes,* 25 November 1958.

Board, and work immediately began on preparation of a draft constitution, which was adopted by the member denominations on 17 March 1960.[28]

The major difference between the new constitution and its predecessor was the inclusion of the Lutheran Church of Australia[29] as one of the constituents, along with the four existing member denominations.[30] Nevertheless, it was not until 1962 that a Lutheran representative, Rev Erich Riedel, was appointed to the Board.[31] Riedel, who had served as an Army chaplain since 1957, was in 1965 posted to the Pacific Islands Regiment, where the large number of Lutherans among the soldiers made the appointment of a Lutheran chaplain desirable. This is significant in that it is the only occasion when a UC chaplain has been given priority on the basis of proportionate representation. Riedel continued to serve in PNG until his retirement in 1972, when his service was recognised with the award of an MBE.[32]

A New Chaplain General

In 1964 at the age of 65 and with his health failing, Brooke resigned his position as Chaplain General. Abbott describes him in his final years in this role as having become the successor to Bishop Riley[33] as the 'patriarch of the Chaplains Department'.[34] He died four years later on 25 September 1968 and was buried in Canberra with full military honours. The Federal Board recorded this tribute to him:

> That this Board place on record its Thanksgiving to God and appreciation for the life and ministry of the Late Chaplain General the Revd. Allen Brooke CBE, ED ... He saw ... service in both Britain and the Middle East, and, on return to Australia

---

28 FUCCB, *Minutes*, 17 March 1960.
29 It then comprised both the Evangelical Lutheran Church and the United Evangelical Lutheran Church.
30 FUCCB, *Minutes*, 25 June 1959.
31 FUCCB, *Minutes*, 30 October 1962.
32 'Chaplain Erich Riedel MBE,' PIB N GIB PIR Association, https://www.soldierspng.com/?_id=4707/.
33 Anglican Chaplain General 1942–1957.
34 Abbott, *In This Sign Conquer*, 414.

was appointed Chaplain General ... then followed a long and distinguished association ... where as Secretary his considerable administrative gifts found expression ... the quality of this service greatly enhanced the stature of the Department ... He was a Chaplain-General of warm heart ... personally interested in his Chaplains ... the appreciated leader, not only of his own group, but in the wider ministry of the Dept ... he exercised a wide influence in the post-war Army.[35]

Brooke's funeral was conducted by his successor, Malcolm McCullough, who served as Chaplain General until August 1981 when a new body, the Religious Advisory Committee to the Services (RACS), replaced the Chaplains General. McCullough's appointment, at first glance, appears strange because the principle of rotational equity, which undergirded the United Churches' *modus operandi* since its inception, suggests that the next Chaplain General should have been a Congregationalist. The original incumbent, Hansen, was a Baptist, and Brooke, though initially from Churches of Christ, transferred to the Baptists in 1946.[36] However, the Federal Board was primarily concerned about appointing the best man for the job, 'giving due but not determinative consideration to the principle of denominational rotation'. Four nominations were received, and McCullough was elected 'by a substantial majority'.[37] He was duly appointed with unanimous support from each of the five member churches,[38] and confirmed in this appointment with effect from 1 July 1964.[39]

There may be a hint of reservation on the part of the Congregational Union, in that its letter supporting McCullough's nomination[40] also included an earlier resolution from its New South Wales Union: 'that steps should be taken to require that the retiring age for the Chaplain General and the Principal Air Chaplain be not more than 60 years'.[41]

---

35 FUCCB, *Minutes*, 11 February 1969.
36 Abbott describes this as a 'studied move to retain the Chaplain General appointment;' Abbott, 413.
37 FUCCB, *Minutes*, 26 March 1963.
38 FUCCB *Minutes*, 16 July 1963.
39 FUCCB, *Minutes*, 11 November 1963.
40 FUCCB *Minutes*, 16 July 1963.
41 Congregational Union of Australia, Letter to FUCCB, 12 June 1963.

FUCCB, however, resolved that it would not change its 'existing practice of these two appointees serving throughout their active ministry'.[42]

Whether this was a move by the Congregational Union to limit McCullough's term to six years,[43] so as to open the door for a future Congregationalist appointment, is not known. Abbott, however, suggests that Congregationalist Hugh Ballard, who had served as an AIF chaplain from 1942 to 1946 and had been decorated for gallantry, was clearly an obvious choice. Ballard had also been Senior Chaplain for Northern Command since 1959 and was well-respected. His widow claimed he was unfairly discriminated against by members of the Board:

> Bob [his middle name] was not selected as Chaplain General (United Churches) as the members of the Congregational Church who were on the board at the time just did not do anything to support Bob's nomination, and allowed remarks to be made that Bob was not a teetotaller. Bob was not a drinker and did much to encourage the men not to drink to excess. Naturally he was very hurt and disappointed.[44]

Abbott also claims that Brooke had groomed McCullough as his successor and had argued, somewhat disingenuously, that it was not inequitable for McCullough, as a Baptist, to succeed him because he [Brooke] had been appointed as a Churches of Christ minister. McCullough's appointment shows that Brooke's opinion prevailed,[45] and the pattern of alternating Baptist and Churches of Christ ministers continued at the Chaplain General and Religious Advisory Committee level until 2019.

A New Principal Air Chaplain

Albiston's retirement as Principal Air Chaplain in 1961, though affecting a much smaller body of chaplains, was proportionately as momentous as Brooke's. Like Brooke, he had taken control of a disparate group of chaplains (an almost non-existent group in his case) on the margins of the

---

42   FUCCB, *Minutes*, 24 March 1960.
43   He was almost 54 at the time of his appointment.
44   H.R. Ballard (Mrs). Letter to Douglas Abbott, 14 August 1986; Abbott, *In This Sign Conquer*, 416.
45   Abbott, *In this Sign Conquer*, 416.

chaplaincy organisation and guided it through a period of exponential expansion and unprecedented challenge. He was influential in shaping it into a body that served effectively, under the most demanding circumstances, and then helping it transition into a largely professional chaplaincy service, able to meet the challenges of the post-war era.

The value of his contribution is well expressed in the following tribute:

> The Federal United Churches Chaplaincy Board, desire to express their thanksgiving to God and their thanksgiving to you for your distinguished service as Principal Air Chaplain 1942–1961, and as the continuing chairman of the Board. Both the extent, and nature, of the leadership demanded of you during the Second World War ... and the consequent quality of your response deserves a higher recognition than is within our power to confer ... we consider such significant and sagacious service has been an outstanding contribution, both in the life of the Kingdom of God and in the life of our Nation.[46]

His successor was Arthur Wilkins (Baptist). In March 1961 the FUCCB recommended his appointment,[47] which took effect on 1 May 1961. He served in that role until his resignation in late 1965 due to failing eyesight.[48] Geoff Crossman[49] (Churches of Christ) was nominated to replace him[50] and was appointed Principal Air Chaplain on 31 December 1966.[51] He held that position until 1985, when he became the UC member of RACS.[52]

Renewed Attempt for Navy Chaplaincy

In 1965 the Federal Board launched another attempt to be included in the Navy's Chaplaincy Branch. Frustrated by the Navy's inaction,

---

46 FUCCB, *Minutes*, 3 May 1964.
47 FUCCB, *Minutes*, 24 March 1961.
48 FUCCB, *Minutes*, 9 November 1965.
49 Son of William Crossman, former United Churches Senior Chaplain in 2nd Military District.
50 FUCCB, *Minutes*, 28 June 1966.
51 FUCCB, *Minutes*, 14 March 1967.
52 Davidson, *Roll and Service Record RAAF Chaplains 1926–1990*.

they determined that if no satisfactory reply was received the member churches would individually approach the Prime Minister.[53] The Secretary of the Navy then advised that the matter had been referred to the Navy Board.[54] The Minister for the Navy rejected the proposal, once again on the basis that there were insufficient adherents of United Churches denominations to warrant such appointments. FUCCB then decided to await the outcome of the pending federal election before deciding on further action.[55] Having later resubmitted their request, the new Minister informed them that the former decision would stand, but offered two reserve chaplaincy positions, one each in Sydney and Melbourne, which the Board agreed to accept.[56] Two more years went by with no reply from the Minister, and so the Board resolved that if no reply was received by its next meeting it would make an official complaint to the Minister of 'discourtesy'.[57] Eventually a reply was received and Robert Ewing (Congregational) and Raymond McKenzie (Churches of Christ) were appointed on 11 and 18 August 1969.[58] Though it was less than what FUCCB had hoped for, it was a significant step forward. The current successor to the original Other Protestant Denominations was now represented in each of the nation's armed services.

Three years later further encouragement came when McCullough was invited to a conference on Navy chaplaincy, called to consider replacing the original 1912 agreement. He advocated for the United Churches to be included in the Protestant Denominations establishment within the Navy Chaplaincy Branch, thereby bringing Navy chaplaincy into line with the Army and Air Force. His advocacy appears to have been well received and he later informed FUCCB that it was likely that United Churches would be recognised as part of the Protestant Denominations Group in its own right.[59]

---

53   FUCCB, *Minutes*, 9 November 1965.
54   FUCCB, *Minutes*, 19 April 1966.
55   FUCCB, *Minutes*, 8 November 1966.
56   FUCCB Minutes, 14 March and 30 May 1967.
57   FUCCB, *Minutes*, 11 Feb 1969.
58   FUCCB, Minutes, 11 November 1969; Lindsay Lockley, 'Congregational Ministers in Australia, 1798–1977', *Camden Theological Library*.
59   FUCCB, *Minutes*, 1 February 1972 and 1 August 1972.

Forward Defence

Coming as it did in the immediate aftermath of Australia's huge commitment to World War II, the years between 1946 and 1972 saw a high demand placed upon its armed services, especially the Army. There was never a time when Australian servicemen and women were not serving somewhere in the South-east Asian region as part of the government's policy of forward defence. Grey refers to those years as being 'characterised by continuous military involvement in South-east Asian wars' and 'also saw the regular forces expand to reach a size exceeded only during the two world wars'.[60]

The British Commonwealth Occupation Force

The first major commitment of Australia's reorganised armed forces was to the British Commonwealth Occupation Force (BCOF), whose first commander was an Australian, Lieutenant General Sir John Northcott. The Australian component was made up of volunteers from the AIF who deployed to Japan in February 1946 to form the 34th Infantry Brigade, later becoming the foundation of the Australian Regular Army.[61] Australia was the BCOF's largest contributor and at its peak included eleven thousand five hundred personnel.[62] Thirteen ships of the RAN and three squadrons from the RAAF also provided support. From 1948, with the departure of other contingents, BCOF became a solely Australian force.

Of the eighteen Army chaplains deployed to Japan[63] three were from United Churches: Alan Farr (Congregational), Harry Clark (Churches of Christ)[64] and Leslie Gomm (Baptist).[65] There were none among the seventeen Air Force chaplains deployed at various times,[66] and none, of course, with the Naval vessels involved.[67]

---

60 Grey, *A Military History of Australia*, 220.
61 In 1948 it was redesignated 1st Brigade and became a major component of the newly formed Australian Regular Army.
62 Grey, *A Military History of Australia*, 202-203.
63 Gladwin, *Captains of the Soul*, 178-79; Abbott, *In this Sign Conquer*, 502.
64 Allen Brooke, *Letter to the Federal United Church Board*, 23 July 1947.
65 VSUCCB, *Minutes*, 28 May 1947.
66 Davidson, *Roll and Service Record RAAF Chaplains 1926-1990*.
67 Rowan Strong, *Chaplains in the Royal Australian Navy, 1912 to the Vietnam War* (Sydney: UNSW Press, 2012), 245.

The United Churches contribution, though tiny in comparison with the scores who deployed during World War II, demonstrates that its chaplaincy was now firmly established in the newly formed Australian Regular Army. The Federal and State Boards supported Farr and Clark with literature, films and a movie projector.[68] Gladwin notes that Gomm, who as PD Senior Chaplain during the Korean War, deployed again to Japan in 1952,[69] brought a new dimension to chaplaincy work through his expertise in radio broadcasting. He had previously developed a reputation in Australia as the 'Radio Padre'. It was a skill he put to good use, broadcasting daily services via the BCOF station WLKS. He also organised a choir that gave live radio performances and performed at various Christmas functions.[70]

High rates of venereal disease among Australian troops and accusations of black marketing caused concern in Canberra, prompting the Minister for the Army, in 1948, to ask Brooke and two other Chaplains General to make an official visit to the BCOF to investigate the reports. This caused concern to the Victorian State Board, whose minutes record: 'Our committee regretted that our Chaplain General Allen Brooke should have agreed to act on the committee of investigation into the moral conduct of our troops in Japan on behalf of the government'.[71] The minutes do not give the reason for its regret, but it may have been influenced by the Returned Services League. The League's President strongly objected to the Chaplains General conducting the investigation, claiming that the accusations of black marketing, gross immorality and poor amenities were baseless. He argued that the Chaplains General, though competent to deal with the matter of amenities, were not competent to address the other issues which required the services of independent, legal experts.[72]

The Chaplains Generals' report, however, found no basis in the accusations, stating that the troops' standard of behaviour and morality was high, the accusations of black-marketing amounted to little more

---

68 VSUCCB, *Minutes*, 12 March 1948, 12 October 1948.
69 VSUCCB, *Minutes*, 25 November 1952.
70 Gladwin, *Captains of the Soul*, 180.
71 VSUCCB, *Minutes*, 12 March 1948.
72 *Daily Advertiser*, 13 March 1948, 2.

than the bartering of some goods, and that amenities provided were praiseworthy.[73] This apparently eased the concerns of the Victorian State Board, and no further mention was made of the matter. It led to renewed efforts to improve amenities, and the introduction of chaplain led moral leadership courses, in support of a renewed campaign against venereal disease[74] which, according to medical statistics, affected 29% of Australians serving with the force. This figure indicates there may have been some substance in the RSL President's claim that the Chaplain Generals were not competent to address such issues.

The Korean War

Australia played a small but significant part in the Korean War, which began on 25 June 1950. The first Australian units to be deployed were the Japan based 77 Squadron RAAF and HMAS *Shoalhaven* and *Bataan*, which were then in Japanese waters. As the crisis worsened the Australian government increased its commitment by sending the 3rd Battalion Royal Australian Regiment (RAR) – then part of the BCOF – to join a British led Commonwealth brigade. The 1st and 2nd Battalions of the RAR also fought in Korea and other ships of the RAN served in Korean waters, including HMAS *Sydney*, Australia's first aircraft carrier. An Australian battalion also remained in Korea after the 1953 Armistice until 1956. Despite its epithet as Australia's 'forgotten war', over seventeen thousand Australians served there, including three hundred and forty who were killed and more than twelve hundred wounded and taken prisoner.

Gladwin, presumably based on Abbott's list of Australian Army chaplains on full-time duty with BCOF and Korea,[75] mentions that twenty-eight Australian Army chaplains served in Korea.[76] However, this number actually equates to the total number that Abbott lists as

---

73 *Scone Advocate*, 30 April 1948, 3.
74 Tom Johnstone, *The Cross of Anzac, Australian Catholic Service Chaplains* (Brisbane: Church Archivists' Press, 2003), 266; Gladwin, *Captains of the Soul*, 182.
75 Abbott, 'In This Sign Conquer,' 502; Abbott's source was N.C. Smith, *Home by Christmas*, (Melbourne: Mostly Unsung, 1990).
76 Gladwin, *Captains of the Soul*, 186.

being on full-time duty in both BCOF and Korea.[77] Seven Air Force[78] and two Navy chaplains were also deployed.[79] The United Churches contribution was three Army chaplains: George McAdam (Baptist), Ray Alexander (Congregationalist) and John Nicholson (Salvation Army).[80] The Salvation Army also provided Red Shield representatives Albert Gray and Edwin Robertson.[81]

The Korean War introduced a new dimension to the Australian experience of war. The North Koreans and Chinese were driven by an ideology that was implacably opposed to western democratic ideals. Furthermore, it was an atheistic ideology bent on the elimination of religious belief: something particularly abhorrent to westerners in the 1950s, when the Church was still considered to be one of the pillars of society.

Nowhere was this abhorrence felt more keenly than in the churches and their representatives in military chaplaincy. They responded enthusiastically to concerns by Government and military leadership that servicemen and women should be morally and spiritually prepared to meet this ideological threat. This led to a call for chaplains to promote Christian leadership courses, disseminate religious literature and lead weekly CO's hours on Christian themes, at which attendance was compulsory for all unit members.[82]

For Australian military chaplains, this signified a new development. In addition to their traditional spiritual and pastoral ministry, they now had a role as promoters of western democratic ideology, which in the 1950s was closely identified with its Judeo-Christian heritage. Similar sentiments were present in the two world wars, particularly in relation to German militarism and Nazism, but not to the extent that chaplains were officially tasked as instructors in ideology. It was a significant addition to their role and raised their profile within the pragmatic military

77 Abbott, *In This Sign Conquer*, 502.
78 Davidson, *Roll and Service Record RAAF Chaplains 1926–1990*.
79 Strong, *Chaplains in the Royal Australian Navy*, 252. Strong notes that the Anglican and Roman Catholic chaplains aboard HMAS *Sydney*, 1951–1952 were, apart from brief visits by chaplains during the Vietnam War, the last Australian Navy chaplains to be deployed on operations until the Gulf War.
80 VSUCCB, *Minutes*, 25 November 1952, 23 February 1954 and 27 April 1954.
81 Gladwin, *Captains of the Soul*, 186-187.
82 VSUCCB Minutes, *20 February 1951*.

organisation. But it also had within it the seeds of future conflicts of interest, the perennial question of to whom and to what is a chaplain responsible: the military and the State, or the Church and God?

National Service Training

In 1951 with the Korean War ablaze, the Australian Government launched Australia's third National Service Training program, requiring all 18-year-old males to receive one hundred and seventy-six days of military training in one of the three services. It led to an increase in the number of chaplaincy positions, both full-time and part-time, as well as three training courses for sixty new chaplains, ten of whom came from United Churches.[83] This figure again represents a significantly higher ratio of UC chaplains than the proportionate system required.

McAdam was the first UC chaplain to work with trainees. His initial posting was to the training depot at Wacol in Queensland. He received high praise from Brooke who spoke of the 'fine job' he was doing there, including getting up to five hundred trainees at his church services, and for the effectiveness of his CO's hour presentations.[84] One year later, he was the first UC chaplain to be posted to Korea.[85]

Moral and Character Training

As previously noted, concern for the development of moral and spiritual resilience in the face of an enemy subject to intense, totalitarian indoctrination, led to a growing emphasis on moral and character training by chaplains in all three services. Six tri-service courses were held in 1956 and Albiston drew special attention to the increasing number of officers attending them. These courses also appear to have had an evangelistic impact. Brooke and Albiston both referred to a number of participants who decided to train for the ministry as a result of the courses and to some 'definite conversions' having taken place.[86]

---

83   VSUCCB Minutes, 20 May 1952.
84   VSUCCB Minutes, 5 June 1951.
85   VSUCCB Minutes, 25 November 1952.
86   VSUCCB Minutes, 26 June 1956, 20 November 1956, 25 November 1958, 25 August 1959.

Another development was the introduction of character guidance courses as part of Army recruit training. In 1959 Brooke reported that all Regular Army recruits at Kapooka would be required to attend a one-week character guidance course presented by chaplains. Six months later he noted that the number of scheduled courses had been doubled to eight with around four hundred recruits at each.[87] The Navy introduced something similar in the 1960s as part of a more general character leadership course, but unlike the Army, attendance was voluntary.[88] Similarly, the Air Force continued to promote voluntary moral leadership courses through its Chaplaincy Branch. This later developed into a compulsory training program for recruits and apprentices. Crossman reported that the syllabus for apprentices would now include eighty sessions of character guidance and citizenship training over two and half years, and eight sessions over ten weeks for recruits.[89]

A major precipitator of this new development in training was the failure of many allied prisoners of war captured by the North Koreans and Chinese to withstand communist indoctrination in what became known as *brainwashing*. Research revealed that a significant factor in those who did resist the process was a developed religious belief system.[90] The original idea seems to go back to a suggestion made by Chaplains General Stewart and Daws, following their visit to Japan in 1947.[91]

Consequently, character guidance became an important part of Australian military chaplaincy, extending to all entry points to each of the three services. The Army Chaplains Department established a dedicated Character Training Team of three chaplains, one each from the Anglican, Roman Catholic and Protestant Denominations, to conduct these courses, supported by other chaplains. The first UC chaplain to be part of the team was Arthur Rothwell (Salvation Army) in 1963.[92] Kenneth

---

87 VSUCCB, *Minutes*, 24 February 1959, 2 June 1959.
88 Strong, *Chaplains in the Royal Australian Navy*, 278.
89 FUCCB, *Minutes*, 11 September 1980.
90 Elizabeth Clayton, 'Re-introducing Spirituality to Character Training in the Royal Australian Navy,' in *Journal of the Australian Naval Institute* (2010), 14-17; E.T. Sabel, 'A History of Character Guidance in the Australian Army', *Australian Defence Force Journal*, No.28 (May/June 1981), 21-22.
91 Report of Chaplains General Stewart and Daws on their Visit to BCOF Japan, National Archives of Australia, MP742/1, 56/1/99.
92 FUCCB, *Minutes*, 16 July 1963.

Jarvis was the next in 1972[93] followed by Ernest Sabel in 1980. Sabel was given the task of reviewing 'the whole field of Character Training and Moral Leadership' including the publication of new resources.[94] He developed a very effective course that covered family relationships, relationships within the Army, spiritual awareness, and making moral choices, particularly in relation to military service. Included in the latter segment was a review of the infamous My Lai Massacre in Vietnam and its ongoing effect on the American soldiers who participated in it.[95]

The Malayan Emergency and Confrontation with Indonesia

In June 1948 Britain declared a state of emergency in Malaya that lasted until July 1960. Australia's involvement began in 1950 with the arrival of transport aircraft and Lincoln bombers. The first Army unit deployed was 2nd Battalion RAR, supported by an artillery battery, in October 1955.[96] Prior to its departure, its chaplain, David Mack (Congregational)[97] joined Chaplain General Brooke in a consecration of the unit colours ceremony.[98] Australian contingents served until the end of the Malayan Emergency, and then stayed on as part of the South East Asia Strategic Reserve, including counter insurgency operations in Borneo during what was called Confrontation with Indonesia in 1965–1966.

Other UC chaplains to serve there included Air Force chaplains George Ashworth (Baptist) in 1965,[99] Kenneth Richardson (Congregational) in 1969,[100] and Army chaplains Denby Holmes in 1968[101] and Rodney Tippett in 1972 (both Churches of Christ).[102]

---

93   FUCCB. *Minutes,* 23 May 1972.
94   FUCCB, *Minutes,* 11 September 1980.
95   Author's reminiscences of serving with Sabel.
96   Grey, *A Military History of Australia,* 222.
97   Congregational Union of NSW Yearbook, 1956, 131.
98   VSUCCB, *Minutes,* 27 September 1955.
99   FUCCB, *Minutes,* 9 November 1965.
100 FUCCB, *Minutes,* 11 Feb 1969.
101 FUCCB, *Minutes,* 13 February 1968.
102 FUCCB, *Minutes,* 28 November 1972.

The Vietnam War

Demanding though they may have been for a nation trying to rebuild its depleted economy, none of these operations matched what it was to experience during its ten-year involvement in Vietnam, which Paul Ham termed 'the last and most prolonged proxy battle of the Cold War'.[103] Compared with the two world wars, the Vietnam conflict was relatively insignificant. The sixty thousand Australians who served there and the five hundred and twenty-one who died are dwarfed by the hundreds of thousands who fought in the two world conflicts and the tens of thousands killed. But the emotional trauma, divisiveness and bitterness generated by Vietnam has had a profound effect on the nation.

The Australian commitment was primarily in Army personnel but included significant numbers from both the Air Force and Navy. It began in 1962 with the arrival of twenty members of the Australian Army Training Team Vietnam and concluded when the platoon guarding the Australian embassy in Saigon was withdrawn in June 1973. At the height of Australian involvement there were around eight and a half thousand Australian troops deployed there.

Of the fifty-five Army chaplains who served in Vietnam, five were from United Churches.[104] The first was Edmond Liddell (Salvation Army) in 1967–1968,[105] followed by Clarence Badcock (Churches of Christ) in 1969–1970,[106] Kenneth Jarvis (Baptist) in 1970,[107] Denby Holmes (Churches of Christ) in late 1970–1971[108] and Ernest Sabel (Lutheran) in July 1971.[109] They were all posted to the Headquarters Australian Force Vietnam, where they also had responsibility for the scattered Australian Army Training Teams.[110] Two others, Rodney Tippett (Churches of Christ, Army) and Glen Brown (Churches of Christ, Air Force), had their deployments cancelled just prior to departure, owing to the newly elected Labor Government's decision to withdraw

---

103 Paul Ham, *Vietnam: The Australian War* (Sydney: Harper Collins, 2007), 6.
104 FUCCB, *Minutes,* 14 November 1967-16 November 1971.
105 FUCCB, *Minutes,* 14 November 1967.
106 FUCCB, *Minutes,* 12 August 1969.
107 FUCCB, *Minutes,* 11 November 1969.
108 FUCCB, *Minutes,* 17 November 1970.
109 FUCCB, *Minutes,* 4 May 1971.
110 Gladwin, *Captains of the Soul,* 203.

all Australian troops.[111] Their number was consistent with the official policy that United Churches provide 9 percent of chaplaincy positions, an indication that the major churches, during that period, were able to meet their required allocations.[112]

One other chaplain who served in Vietnam was Donald Woodland (Salvation Army). His experience highlights the usually collegial but sometimes strained relationship between chaplains and philanthropic representatives. In 1969 while serving as a CMF chaplain the Salvation Army sent him to Vietnam as its Red Shield Welfare Service representative. This again raised the question of demarcation between chaplains and philanthropic representatives, requiring him to resign as a chaplain. He reported having been visited by Chaplain General McCullough who he said 'left me in no doubt that when I went to Vietnam, I was in no way to carry out chaplaincy duties because I was only a philanthropic Representative'.[113] In his autobiography[114] Woodland says that chaplains 'in those days, showed a fair amount of animosity towards the Sally Man,'[115] which may indicate that memories of World War II controversies, previously mentioned, were still present. Nevertheless, he developed a close friendship with the battalion's official chaplain[116] and they worked in close cooperation, often travelling together in Woodland's Red Shield Services vehicle and using his *Hop In Centre*[117] as a place where the chaplain could mingle with the troops.[118] Despite the official lines of demarcation the soldiers, including the CO, referred to Woodland as 'Padre:'[119] another indication that troops base their assessment of who is or isn't a chaplain on function rather than appointment.

---

111 FUCCB, *Minutes*, 16 November 1971. Tippett reported to the writer that he was actually waiting at the Personnel Depot, ready to depart when news came through that all Australian forces were to be withdrawn from Vietnam.
112 55 Army chaplains were deployed: Ham, *Vietnam*, 428; Gladwin, *Captains of the Soul*, 203. Ten Air Force chaplains were also deployed: Davidson, *Sky Pilot*, 10-14; Roll and Service Record RAAF Chaplains 1926–1990.
113 Don Woodland, email correspondence to author, 26 May 2021.
114 Don Woodland, *Picking Up The Pieces* (Sydney: Macmillan, 2006), 42.
115 Army slang for Salvation Army Red Shield Services representative.
116 Anglican chaplain Stan Hessey.
117 The Red Shield's amenities facility characterised by its famous red kangaroo logo.
118 Woodland, email.
119 Woodland, *Picking Up The Pieces*, 46.

On his return to Australia Woodland was reinstated as a chaplain.[120]

Denby Holmes also encountered some reticence, reflecting that initially some chaplains were 'a bit unsure about having a Churches of Christ chaplain'. One Roman Catholic told him he did not believe that the Churches of Christ were a true church, echoes of 1915 and Archbishop Riley. Over time this changed, and the nature of operational chaplaincy again resulted in a flowering of mutual respect and practical ecumenism. Holmes developed a very close friendship with the Roman Catholic padre and reported that he would regularly attend Holmes' weekly Sunday services to listen to his sermons. Those services were strictly non-denominational, and usually had around thirty soldiers present, including the base commander who 'always sat in the front row'. Holmes described his ministry as mostly providing pastoral support to troubled soldiers, undergirded by an evangelistic desire to evoke faith as a soldier's greatest resource to face the traumas of war.[121] It also included the trauma of ministering to badly wounded men, especially those lying on their own or in a state of shock.[122]

As in previous conflicts chaplains, in addition to their explicit religious roles, were counsellors and confidants to the troops, providing a listening ear and understanding heart to those who were anxious, traumatised and grief-stricken. As embedded members of their units, they were the first responders in providing psychological and spiritual care for those affected by traumatic events.[123] Brigadier Philip McNamara, as a young platoon commander, never forgot the beneficial effects of a chaplain's presence when his much-respected platoon sergeant was killed. The first person to alight from the evacuation helicopter, he reported, was the battalion padre, who then spent several days with them, quietly helping soldiers deal with their loss as the patrol continued.[124]

The ethical and moral controversies that surrounded the Vietnam War did raise problems for chaplains when people associated their

---

120 Woodland, email.
121 Denby Holmes, Interviews with Author, 16 April 2021, 30 October 2022.
122 Keyes, Don and Williams, Andrew. *Say a Prayer for Me, The Chaplains of the Vietnam War* [documentary film], Sydney Headquarters Training Command Australian Army, 1995.
123 Australian military psychologists were not deployed to operational areas until the Rwanda operation in 1995.
124 Philip McNamara, conversation with Robert Smith, 1 November 2022.

ministry with the promotion of the war effort. There is evidence that many American soldiers lost faith in chaplains, some of whom saw their role to be legitimisers of the American war effort. Charles R. Figley's research reveals the bitterness that many veterans felt towards chaplains who shared this uncritical faith in the moral rightness of American policy, assuring troops that they had been doing God's will, while the soldiers themselves knew that much of what they had done was the very opposite.[125] Jacqueline Whitt concludes that: 'The evidence is unsettling. In the face of atrocities and war crimes, particularly those committed by Americans, chaplains appeared to do little in response.'[126]

There is little evidence of this by Australian chaplains. Even so, national serviceman Andrew Treffry, who served in Vietnam 1968–1969, wrote about how, following a lecture from a chaplain, he felt confused at the presence of a man of God 'amongst a group of trained killers'.[127] It is more likely, however, that most Australian veterans would echo the words of another national serviceman, Barry Heard, who recalls: 'There were good people [there] called padres' and 'the Sallyman is one of my fondest memories'.[128] Paul Ham probably spoke for most when he wrote: 'Many ... soldiers, even those without faith, found the chaplain's presence a kind of hope in hell'.[129] Nevertheless, the Vietnam War confronted Australian chaplains with moral ambiguities greater than any faced by their predecessors in the two world wars.

The End of an Era and the Beginning of a New

Geoffrey Blainey traces the radical shift in social structures that occurred during the Vietnam War to the turmoil that followed World War II.[130] Throughout this period, Australia's armed services were almost continuously involved in operations to stem the tide of Communist expansion.

---

125 Charles Figley, *Stress Disorders Among Vietnam Veterans: Theory, Research and Treatment* (New York and London: Brummer-Routledge, 1978), 220.
126 Jacqueline Witt, 'Faith Under Fire: Military Chaplains and the Morality of War'.
127 Papers of A. Treffry, AWM PR00032.
128 Barry Heard, *Well Done Those Men* (Melbourne: Scribe, 2005), 105, 112.
129 Ham, *Vietnam*, 428-429.
130 Geoffrey Blainey, *A Short History of the 20th Century*, (Melbourne: Viking, 2005), 408.

For the Army, except for the two periods of national service, it was also a period of change from a predominantly part-time force to a professional regular army supported by the citizens force. The same process gradually evolved within the chaplaincy organisations of both the Army and Air Force. In 1970, following the example of the other denominational groups, FUCC resolved to pursue the matter of permanent commissions[131] and, at Crossman's request, agreed to 'avail itself of the opportunity afforded (in Air Force)'.[132] The first UC chaplain to receive a permanent commission was Glen Brown (Churches of Christ) on 13 September 1970.[133] Later that year, the Board agreed to a proposal by McCullough that permanent commissions be sought for Army chaplains who had served full-time for at least five years and were prepared to retire at the age of 50.[134] Consequently, in 1973 Denby Holmes became the first UC Army chaplain to be given a permanent commission.[135]

During this period the United Churches, which now included the Lutheran Church, consolidated its position as an established provider of chaplains for the Army and Air Force, and finally achieved recognition in the Navy.[136] Through its Chaplains General and Principal Air Chaplains, especially Brooke and Albiston, it gained significant credibility at the highest levels of Army and Air Force chaplaincy, and also through its chaplains who served at home and on operations. It had stood fast amidst the heavy demands of the post-war era and was now ready to face the challenge of a very different environment.

---

131   RAN full-time appointments had been permanent since 1952; Rowan Strong, *Chaplains in the Royal Australian Navy, 1912 to the Vietnam War* (Sydney: UNSW Press, 2012), 255.
132   FUCCB, *Minutes,* 10 February 1970.
133   FUCCB, *Minutes,* 11 August 1970.
134   FUCCB, *Minutes,* 17 November 1970.
135   FUCCB, *Minutes,* 2 May 1973.
136   Except for the two Reserve appointments in 1969.

CHAPTER EIGHT

Post-Vietnam: The Emerging Stone

THE DECADE THAT FOLLOWED the Vietnam War was in many respects the most difficult ever for Australia's armed services. The erosion of public esteem for the military was keenly felt, and for veterans was a source of bitterness that compounded traumatic memories.¹ Peter Firkins said of them:

> They had fought well and courageously, upholding all the finest traditions so bitterly won by Australians at war ... Their bearing and discipline had been an example to all who served in Vietnam with none of the drug and related problems that so affected ... the US Army ... [but] [f]or servicemen who had performed so well they were at times shabbily treated ... Federal governments which were responsible for sending them, did little to defend their honour ... For the first time in Australia's history, its people had deserted its fighting men.²

Defence was not a high priority for the newly elected Labor government, which saw 'no perceivable threat' to Australia for at least ten

---

1 Paul Ham, *Vietnam, The Australian War* (Sydney: Harper Collins, 2007), 564.
2 Peter Firkins, *The Australians in Nine Wars, From Waikato to Long Tan* (Sydney: Pan, 1982), 483-484.

years,³ and public antipathy towards the military, though far from universal, severely undermined the 'Anzac Legend', which had been central to Australia's national self-image since 1915. Serving members were told not to wear their uniforms in public lest they be spat at – or worse.⁴ Denby Holmes, reflecting on his return from Vietnam, described this as 'a very sad experience'. He was proud of the uniform, and for the nation to be ashamed of it, having sent him to war wearing it, was emotionally devastating.⁵ The damage to morale continued well into the following decade, as was evident in a comment to the author in 1987 by one of the Army's most senior leaders. Reflecting on what he perceived as a diminishing belief in the military by both the public and the government, he admitted: 'And now we are not sure we even believe in ourselves'.

It was not until 3 October of that year, when twenty-two thousand Vietnam veterans marched through the streets of Sydney in the Australian Vietnam Forces Welcome Home Parade, that the wounds began to heal. By then a more realistic appraisal of the conduct of Australian troops had emerged, along with a desire to rectify the injustice they had suffered. Prime Minister Bob Hawke foreshadowed it in a speech to the Returned Services League:

> I firmly believe that the October parade will be the culmination of a long process of reconciliation and community acceptance of its obligations to the veterans of Vietnam ... But whatever our individual views on the merits of Australian involvement, we must equally acknowledge the commitment, courage and integrity of our armed forces who served in Vietnam.⁶

Chaplain Veterans

No less than other Vietnam veterans, chaplains shared in the war's bitter legacy. For some the most painful experience was the reception they received from their fellow clergy. Lester Thompson (Anglican) reflected

---

3   *Canberra Times*, 9 October 1975, 6.
4   Author's personal recollection; Michael Gladwin, *Captains of the Soul. A History of Australian Army Chaplains* (Sydney: Big Sky, 2013), 229.
5   Interview with Denby Holmes, 14 November 2022.
6   Bob Hawke, Address to the RSL Conference, August 1987.

sadly on how shattered he felt after visiting his bishop following his return from Vietnam. Traumatised and confused, he went seeking fellowship and pastoral support but instead was harangued about the evil of what he had been involved in.[7] Similarly, Roy Bedford (Methodist) described the hostility he encountered from fellow clergy, and the wounds it created. He confessed that even years later he found it hard to accept and difficult to forgive.[8] Howard Dillon (Anglican) described it as 'a mortal blow' and directed his criticism mostly at 'those community leaders who ... really piddled on us from a great height'.[9]

The five UC chaplains appear to have been spared that added trauma. Their recorded memories indicate they received little if any response from their denominational leadership, neither praise nor criticism, just 'business as usual'. Holmes reflected that unlike some of his colleagues from other denominations he received no criticism from his denomination, but neither was he given any pastoral support despite what he had lived through. On the other hand, he often felt supported by ordinary church members, particularly when invited to preach in local churches where he found 'the responses were always positive'.[10] Ern Sabel's experience was similar. While friends and local clergy colleagues welcomed him home and showed interest and support, he received 'no contact or interest from church leaders and general clergy'. His biggest disappointment was 'the lack of interest from the leadership within the Chaplains Department, who gave no thought to the need for debriefing and learning from the experiences of returned chaplains'. It corresponded to the lack of training he'd received before being sent to war, causing him to conclude: 'At that time it was assumed that ordination produced all that was needed to be a chaplain at war'.[11]

The failure of the churches and Chaplains General in this respect draws attention to the overall lack of understanding of post-traumatic stress. It was generally assumed that chaplains returning from active

---

7 Author's recollection of a conversation with Lester Thompson in 1982, in which he expressed his continuing trauma over the way he was received back by his church.
8 Don Keyes and Andrew Williams, *Say a Prayer for Me, The Chaplains of the Vietnam War* [documentary film], Sydney Headquarters Training Command Australian Army, 1995.
9 'Howard Dillon (Australian Army) Army Chaplain,' Anzac Portal.
10 Interview with Denby Holmes, 14 November 2022.
11 Ernest Sabel, email to Robert Smith, 17 November 2022.

service would just get on with life and that their own spiritual resources would be sufficient to deal with any lingering difficulties. This was also a sad reflection of how quickly the Chaplains General had forgotten the lessons from their own wartime experiences and may reflect the condescending attitude of many World War II veterans towards those who fought in Vietnam. This was even true of the Returned Services League which as Major General Gration recalls 'Didn't believe it [Vietnam] was a real war'.[12]

Don Woodland also felt keenly the lack of pastoral support. It was only after his return home that he realised how traumatic Vietnam had been. 'My nerves were raw ... and even to drive at night and see the reflector tags ... was enough to send me into a mild panic attack'. It placed enormous stress on his marriage, and he and his wife were left to deal with it as best they could. He confessed: 'We had some rough days for two, three, four years, mainly because we had absolutely no support from the military or understanding or support from the Salvation Army'.[13] Nevertheless he, like the five United Churches chaplains, was spared the hostility experienced by colleagues from other denominations. This may reflect the innate conservatism of most of their parent denominations, which tended to be implacably opposed to Communism and more ambivalent about opposition to the war than the more 'liberal' major churches.

The Churches and the War

In order to understand how this complex situation developed it is important to trace the changing attitude to the war within the Church and community. Despite there being a strong core of conservative pro-war sentiment in the churches generally, as the anti-war forces in the community gradually became stronger, so did their supporters in the churches. Ian Breward agrees, reflecting that as early as 1965 two Anglican Archbishops and seven diocesan bishops were urging Prime Minister Menzies to seek a settlement to the conflict. It was, however,

12   Ham, *Vietnam*, 565.
13   Don Woodland, *Picking Up The Pieces* (Sydney: Macmillan, 2006), 67, 69.

the 'extensive television coverage of the napalm bombing of civilians, the use of defoliants, atrocities and the unattractive character of the [South Vietnamese] regime' that gradually changed the perception of the conflict as a 'just war'. It justified the strident criticism of Methodist notables like Rev Alan Walker and Dr Allan Loy, who had opposed it from the start.[14] Even so, this did not always meet with the approval of the laity, as when the Principal of Newington College, Rev Douglas Trathen, was dismissed over his anti-war statements.[15]

Apart from *Pax Christi*,[16] the Roman Catholic Church tended to see the Vietnam conflict as a 'just war' influenced, no doubt, by the large number of Roman Catholics in South Vietnam who faced a very uncertain future under communist rule. Similarly, the conservative-evangelical Anglican Diocese of Sydney tended, at least initially, to oppose the anti-war stance of other Anglican leaders. Archbishop Hugh Gough's visit to Vietnam in 1965 to demonstrate his support for the war was a notable example of this. Even more striking was the 1967 incident in which Rev Broughton Knox, Principal of Moore Theological College and Chairman of the NSW Council of Churches, had a live broadcast on Radio 2CH (whose licence was held by the Council of Churches) terminated when Rev Alan Walker urged support for Senate candidates who would pledge to end the Vietnam conflict.[17]

Opposition to the war appears to have been more muted within the United Churches but did grow as the war dragged on. Stuart Piggin and Robert Linder mention that it was 'particularly intense among Victorian evangelicals ... especially among the Baptists,[18] and Ros Otzen cites gradual opposition to the Vietnam War as a key indicator of changing attitudes among young Baptists.[19] Similarly, the Congregational Union appears to have become increasingly uneasy, as seen in Rev. J.W. Brookfield's 1969 presidential address:

---

14   Ian Breward, *A History of the Australian Churches* (Sydney: Allen & Unwin, 1993), 185.
15   Douglas Arthur Trathen, Papers, 1939–1972, Canberra, NLA, MS 10546.
16   A Roman Catholic peace movement.
17   Stuart Piggin and Robert Linder, *Attending to the National Soul* (Melbourne: Monash University Publishing, 2020), 333-334.
18   Piggin and Linder, *Attending to the National Soul,* 334-335.
19   Ros Otzen, 'Major Trends Among Victorian Baptists 1939–1965,' *Our Yesterdays,* Volume 16 (2008), 22.

> How many of us, despite our lip-service to the Prince of Peace, are prepared to demand that our Government take immediate action to end the war in Vietnam, for the sake of saving the Vietnamese people any further futile suffering? ... still imagining that by killing people we can kill a philosophy.[20]

Yet apart from this, there is little if any mention of the war in the Congregational Union Yearbooks between 1965 and 1972, although it is interesting to note that contrary to previous wars no Congregationalist chaplains were deployed to Vietnam. As for the Churches of Christ, Salvation Army and Lutheran Church, apart from occasional protests, such as Dr Desmond Crowley's article opposing the use of conscripts in Vietnam,[21] there appears to have been little organised opposition.

Theological conservatism may account for this reticence. David Mislin argues that the division between liberal Protestant leaders and more conservative churchgoers in America widened in the Vietnam era, and that many conservatives, like the editor of *Christianity Today*, Carl F. Henry, believed the war to be morally defensible, and opposition to be unpatriotic.[22] The latter part of the war did see the emergence of a changing attitude within the United Churches, but not to the extent of focusing its anger on its returned chaplains.

Advent of the Australian Defence Force

In military organisation, training for operations and actual operational involvement is followed by a third phase known as reconstitution: restoring it to a desired level of combat effectiveness. The quarter century that followed the Vietnam War was, in many ways, a period of reconstitution for Australia's armed services and for its chaplaincy bodies. The move towards a professional military chaplaincy in an increasingly professional military environment accelerated during this period, and by

---

20 J.W. Brookfield, 'President's Address,' *Congregational Union of New South Wales Yearbook 1969*, 17-18.
21 *Australian Christian*, 7 May 1966, 8.
22 David Mislin, 'How Vietnam War Protests Accelerated the Rise of the Christian Right,' *Smithsonian Magazine* (3 May 2018).

the end of the century was complete. This was particularly noticeable in the Army, which in 1972 began to reorganise from geographical to functional commands,[23] but was also evident in Air Force Chaplaincy. The Navy, on the other hand, had traditionally been served primarily by full-time chaplains, supported by reservists.[24]

The main catalyst for this professionalisation was the creation in 1976 of the Australian Defence Force (ADF) as the overarching military organisation tasked with the defence of Australia and comprising the three armed services. Responsibility for its creation was delegated to the formidable Secretary of the Department of Defence, Sir Arthur Tange who, as Jeffrey Grey asserts, 'is more closely identified with this period than anyone else'.[25] It was Tange who in 1981 decided that a new governance body for chaplaincy was needed to replace the Conference of Chaplains General, and named it the Religious Advisory Committee to the Services (RACS).[26] Its inaugural meeting took place on 23 September 1981,[27] and its scope was wider than its predecessors covering the whole ADF, but without administrative responsibility, which was delegated to a newly formed group of Principal Chaplains in each service, one from each of the Anglican, Roman Catholic and Protestant Denominations.[28]

RACS was set up as a civilian body and its members, though not entitled to military rank, were given two-star status[29] and the Principal Chaplains one-star rank.[30] The respective roles of each body were set out in a Memorandum of Arrangements first signed in 1981. Essentially, whereas the Principal Chaplains would administer the chaplaincy organisations, RACS became the official interface between the Churches and the ADF. Its role included advising the Chief of Defence on policy related to religious and spiritual well-being of all members

---

23 Field Force, Training and Logistics.
24 Rowan Strong, *Chaplains in the Royal Australian Navy 1912 to the Vietnam War* (Sydney: UNSW Press, 2012), 255-256.
25 Jeffrey Grey, *A Military History of Australia* (Melbourne: Cambridge University Press, 2008), 256.
26 Religious Advisory Committee to the Services, NAA A1114, 1981/49 PART 1; FUCCB, *Minutes,* 29 January 1981.
27 Religious Advisory Committee to the Services, *Minutes,* 23 September 1981.
28 One Principal Chaplain to serve full-time and the other two part-time on a rotating basis.
29 Equivalent to Major General and equivalents.
30 Brigadier and equivalents.

of the Defence Force, advising service chiefs on the appointment of Principal Chaplains, formulating religious policy for implementation by Principal Chaplains, and authorising forms of worship for use in ecumenical services and on national occasions. Individual members of RACS had pastoral oversight of all chaplains within their denominational groups and became the authorising agents for all new chaplains within those groups. They provided continuing endorsement, exercised ecclesiastical discipline, monitored chaplains' professional development and facilitated their return to civilian ministry.[31]

Appointment of Full-time Chaplains in Navy

In 1974 the Navy Board finally agreed to the appointment of a full-time chaplain from the United Churches,[32] and FUCCB acknowledged Malcolm McCullough's successful advocacy in a special minute thanking him for his 'continuing presentation of the case for United Churches representation within Navy and upon the successful outcome'. It was an historic milestone in the journey towards equal participation with the major churches.[33] Two months later a special Board meeting nominated Kenneth Jarvis (Baptist) to be 'The Board's first career chaplain to RAN' and arranged for him to be transferred from the Army.[34]

Jarvis proved to be an excellent choice and in 1981 was joined by a second full-time chaplain, Brian Daniel (Baptist).[35] His initial posting was to the Navy's training establishment at HMAS *Leeuwin* in Western Australia, which he described as a 'fascinating challenge,' where the '700 lads responded remarkably well to my ministry'. He referred particularly to ten voluntary weekend programs when he and Roman Catholic chaplain Kevin Ryan took thirty-five trainees off-base for a weekend experience of discussion groups and visiting speakers, describing their

---

31 Memorandum of Arrangements between the Commonwealth of Australia and the Heads of Churches Representatives, 31 March 1981; RACS, *Minutes*, 11 December 1990.
32 Letters from Naval Board to Federal United Churches Chaplaincy Board, 15 January 1974, 18 April 1974 and 10 June 1974, Baptist Union of Victoria Archives.
33 FUCCB, *Minutes*, 27 August 1974.
34 FUCCB, *Minutes*, 29 October 1974.
35 FUCCB, *Minutes*, 13 October 1981.

response as 'exceptional'. In addition to much individual pastoral support, he was heavily involved in providing religious instruction to the trainees and had four courses running continuously. Despite his regret at leaving HMAS *Leeuwin*, he reported looking forward to his next posting, seeing it as 'another milestone, as the United Churches will have their first ever Chaplain to the Fleet'.[36] Federal Board minutes then report him being chaplain for 'Small Ships,' which Jarvis explained as involving pastoral care of the destroyer squadrons and spending much time at sea, moving from ship to ship.[37]

Six months later, while preparing to sail to the United States aboard HMAS *Vampire* to participate in the American Bicentenary celebrations, he wrote of the difficulties associated with being chaplain to the Fleet.

> I find the Fleet ministry a pretty frustrating one ... it's just extremely difficult to make contacts ... Nevertheless, I can see the very real value of such a ministry ... I guess one has to be a part of it to really experience the pressure a sailor is under ... I was amazed at the pressure a Ship's crew is put under in operational conditions.[38]

It was the beginning of a distinguished service that in 1984 saw him become the first ever Australian Navy chaplain to be enrolled in the Navy's Staff Officer Course, and three years later he become the Navy Principal Chaplain (Protestant Denominations).[39]

United Churches Principal Chaplains

Jarvis was not the first UC chaplain to be advanced to Principal Chaplain. Mention has already been made of Walter Albiston's appointment as Principal Air Chaplain in 1955. He was followed by Arthur Wilkins in 1961, and then by Geoff Crossman in 1966. Prior to the changes brought about by Sir Arthur Tange in 1981, the five Principal

---

36 Ken Jarvis, letter to the FUCCB, 3 November 1975.
37 FUCCB, Minutes, 18 November 1975, 24 February 1976.
38 Ken Jarvis, letter to David Griffiths, 24 May 1976.
39 FUCCB Minutes, 11 September 1984, 10 November 1987.

Air Chaplains had served primarily as Principal Chaplains for their respective denominational groups. However, their role now became functional as well as denominational, with staff responsibilities for all chaplaincy work within the Air Force.[40] Their number was reduced to three – one Anglican, one Roman Catholic and one Protestant Denominations, each serving full-time. Crossman, who had been appointed Secretary of the Board of Chaplains in 1978,[41] became the full-time Protestant Principal Air Chaplain until his retirement in 1985, having served thirty-two years.[42]

The Army Principal Chaplains Committee comprised one full-time Regular Army Principal Chaplain, who acted as chairman, and two part-time Army Reservist Principal Chaplains, one each from the other two denominational groups. David Griffiths (Baptist) was appointed Protestant Denominations Principal Chaplain in July 1981[43] and served part-time until his retirement in early 1988. He was succeeded in February of that year by Ern Sabel as full-time member of the Army Principal Chaplains Committee.[44] In March 1991 Sabel retired from the Regular Army, relinquishing the chairman's role but remaining a reserve member of the Principal Chaplains Committee for the next three years.[45]

The Navy was the only service that did not appoint a chaplain from United Churches as its first Protestant Principal Chaplain. Even so, Jarvis' appointment on 2 November 1987,[46] a mere six years after the structural changes that brought the Principal Chaplains Committees into being, and only 13 years after he had become the first UC chaplain in Navy, speaks highly of his ability. The fact that UC chaplains dominated Protestant chaplaincy administration in all three services for much of the 1980s also reflects their emergence as a growing influence in Australian military chaplaincy.

---

40 Peter Davidson, *Sky Pilot: A History of Chaplaincy in the RAAF 1926–1990* (Canberra: Directorate of Departmental Publications, Department of Defence), 14-16.
41 Air Vice Marshall Parker, letter to Chaplain General McCulloch, 26 January 1978, Baptist Union of Victoria Archives; FUCCB, *Minutes*, 25 May 1978.
42 FUCCB, *Minutes*, 14 May 1985.
43 RACS, *Minutes*, 9 December 1981.
44 RACS, *Minutes*, 9 December 1987.
45 FUCCB, *Minutes*, 10 November 1987; 9 April 1991.
46 FUCCB, *Minutes*, 10 November 1987.

Vale the Congregational Union

When the Uniting Church of Australia came into being on 22 June 1977, the Congregational Union of Australia ceased to exist as a separate body. Consequently, its involvement in ADF chaplaincy was transferred from the United Churches to the Uniting Church, ending a relationship that had existed since before World War I. This historic event, however, seems to have gone almost unnoticed in the minutes of the FUCCB. Apart from a brief mention, three years earlier, about making 'enquiries to the Congregational Union' concerning the future relationship between it and the FUCCB, and a reference to the need to discuss 'the implications' of the formation of the Uniting Church in Australia',[47] there is no mention of its departure in the Board's minutes. After such a long and vital involvement, one would have expected there to have been appropriate motions of appreciation and good wishes. However, it is possible that FUCCB anticipated a continuing Congregational presence through those congregations that chose not to join the Uniting Church.

Even so, it was five years before any discussion took place. In 1982 McCullough reported having met with representatives of the Fellowship of Congregational Churches[48] who expressed the desire to continue the Congregational Church's association with FUCCB. The Board consequently agreed that the Congregational Fellowship should make a formal application in writing, which it did, and in March 1984 was accepted as a member.[49] However, since that decision was made, The Fellowship of Congregational Churches has never appointed a representative to FUCCB and has only ever nominated one chaplain: Martin Sharteris, who was appointed to the Army Reserve in 1987.[50] This may well be a reflection of the difficulties that confronted that rather small group of independent congregations, as they attempted to reestablish their national profile, than a lack of interest in military chaplaincy.

---

47   FUCCB, *Minutes*, 28 May 1974: 9 August 1977.
48   Which, along with the more liberal Congregational Federation of Australia, represented the congregations which chose not to become part of the Uniting Church.
49   FUCCB, *Minutes*, 14 September 1982; 20 March 1984.
50   RACS, *Minutes*, 2-3 September 1987.

Interest from Other Denominations

The Seventh Day Adventist Church (SDA) enquired about joining FUCCB in 1982 and was advised to write to the Minister for Defence seeking recognition within the terms of the Memorandum of Arrangements between Defence and the Churches.[51] It appears not to have done this but then made another approach to FUCCB in 1987.[52] The Board declined its request but agreed that McCullough would represent SDA interests at the Religious Advisory Committee to the Services.[53] Following this there was no further interest shown until 2017 when another application was made and formally approved the following year.[54]

The next expression of interest came in 1988 when the Assemblies of God (AOG) wrote to FUCCB seeking membership.[55] The December meeting noted that the Assemblies of God, though the largest, were just one of many Pentecostal denominations in Australia, and that to admit them would 'require a variation of the Memorandum of Arrangements'. Concern was also expressed about the possibility that serving chaplains might 'experience difficulties over proselytising activities' by Pentecostal chaplains. The Board consequently resolved 'not to alter the constituency of the Federal Uniting Churches Chaplaincy Board at this time'.[56]

It is interesting that no such concern over proselytising was mentioned in relation to the SDA enquiry. It indicates a level of unease felt towards Pentecostalism that permeated much of mainstream Christianity at that time, often focussed on the perceived 'emotionalism' of Pentecostal worship, as well as the scandals associated with certain high profile Pentecostal ministers. It may also reflect a degree of envy of Pentecostal church growth compared with diminishing numbers in the more traditional churches.

The matter was raised again in 1991 in a letter from the Australian

---

51 FUCCB, *Minutes*, 23 March 1982.
52 R.L. Coombe, letter to G. Crossman, 11 November 1987, Baptist Union of Victoria Archives.
53 Letter from David Griffiths to R.L. Coombe, 2 June 1988, Baptist Union of Victoria Archives.
54 FUCCB, *Minutes*, 19 March 2017; 9 March 2018.
55 Assemblies of God in Australia, letter to FUCCB, 27 June 1988, Baptist Union of Victoria Archives.
56 FUCCB, *Minutes*, 7 December 1989.

Pentecostal Ministers Fellowship, which represented ten Pentecostal groups within Australia.[57] FUCCB had by then softened its approach, and the Baptist Union openly stated that it had no objection to their involvement, seeing them as 'within the Protestant tradition'.[58] The Salvation Army, however, was still unsure, and thought the Pentecostal churches had to be seen 'as a new and separate grouping'.[59] Consequently, FUCCB agreed to offer the Assemblies of God a consultative role on the Board and indicated it would represent their interests in Defence. It also agreed to ask RACS to reconsider its structure so as to allow further representation in the light of Australia's multi-religious nature.[60]

RACS, however, was not prepared to accept the proposal, arguing that 'Churches of the Pentecostal order do not fit comfortably within the Federal United Churches Chaplaincy Board',[61] a somewhat disingenuous response that failed to account for the differences that have always been evident within the United Churches. Furthermore, there were by then enough Pentecostal adherents in the Defence Force to meet the requirements of the Memorandum of Arrangements for nominating candidates for chaplaincy.[62]

Nevertheless, it is clear that FUCCB had moved beyond the reticence of the major churches in its awareness of the changing nature of religious life in Australia, and its implications in relation to the diminishing pool of potential new chaplains. Confirmation of this came in 1993 when the Assemblies of God formally recommended Queensland based Pastor Michael Alcock for Army Reserve chaplaincy. This time there was no hesitation based on perceived problems about his ability to adapt to the culture of military chaplaincy. He was interviewed by FUCCB's chairman, Rev David Griffiths, considered to be acceptable

---

57 Australian Pentecostal Ministers Fellowship, letter to FUCCB, 6 June 1991, Baptist Union of Victoria Archives.
58 Baptist Union of Australia, letter to FUCCB, 31 May 1992, Baptist Union of Victoria Archives.
59 Salvation Army Southern Territory, letter to FUCCB, 21 May 1992, Baptist Union of Victoria Archives.
60 FUCCB, *Minutes,* 29 June 1992.
61 Ralph Estherby, email to Robert Smith, 4 August 2022.
62 Memorandum of Arrangements between the Commonwealth of Australia and the Heads of Churches Representatives, Revised 1991 and 2004, 2 December 2004.

and appointed an Army Reserve chaplain on 17 July 1994.[63] The Board, however, also agreed that 'no further [AOG] appointments [would] be made at this stage'.[64]

Despite the recalcitrance of RACS, the emergence of Pentecostalism was being widely recognised. Harvard theologian Harvey Cox, in a research project that took him to four continents, documented its phenomenal worldwide growth, then estimated to be twenty million new members each year, culminating in a worldwide membership of four hundred and ten million.[65] Similarly, in Australia, 'From 1961 to 2011, the Pentecostal denominations grew from about 16,000 people in Australia to perhaps around 250,000 ... from a fringe movement to one of the largest denominational groups'.[66]

The response of RACS revealed its failure to see that its member churches no longer had a monopoly on the religious life of the nation and echoed the dismissiveness of the original Chaplains General towards the Other Protestant Denominations. It was fortunate that FUCCB did not share that reticence.

From Class System to Divisional System

During 1989–1990 the Defence Force Remuneration Tribunal approved a new chaplaincy classification structure, enabling chaplains to be both advanced and paid according to skills and ability, rather than seniority and time served. Sabel, supported by his equivalents in Air Force and Navy, was largely instrumental in facilitating this. The change meant that instead of the existing systems of relative rank, chaplains would be classified within five 'divisions' or levels of seniority corresponding to relative ranks within the three services – Divisions 1 to 5 relating to captain, major, lieutenant colonel, colonel and brigadier and their equivalents. A new emphasis was placed on competency-based training and in 2003

---

63 FUCCB, *Minutes,* 9 November 1993; 18 May 1994.
64 FUCCB, *Minutes,* 30 October 1996.
65 Harvey Cox, *Fire from Heaven* (London: Cassell, 1996), xv.
66 Philip Hughes, *Charting the Faith of Australians: Thirty Years in the Christian Research Association* (Melbourne: Christian Research Association, 2016), 35.

led to the creation of the tri-service Defence Force Chaplains College.[67] It meant that the traditional 'committee system', whereby chaplains worked collegially in committees representing the three major religious groups: Anglican, Roman Catholic and Protestant Denominations would be largely superseded by an administrative system based on a chaplain's division classification. Since advancement was no longer to be tied to denominational quotas, it was a further move towards generic rather than denominational chaplaincy.

This corresponded with a major restructure of Army chaplaincy that Sabel initiated and implemented. The Navy quickly followed with a similar restructure and soon after the Air Force did the same. In August 1990 Sabel reported that the restructure had been well received by Army chaplains,[68] and Jarvis announced that Navy, as part of its restructure, had appointed two chaplains at Division 3 level, bringing it into line with Army and Air Force.[69]

Sabel also reported that all regular army chaplaincy positions were now filled. Sadly, this was short-lived. As the decade progressed so did a chronic shortage of Anglican, Roman Catholic, and Uniting Church chaplains. The overwhelming problem was their lack of younger clergy who were interested in military chaplaincy, suited to service life, and physically able to meet the medical and physical fitness requirements. This was especially concerning for the Roman Catholics whose ever-diminishing number of priests had by far the oldest average age of all the churches.

In contrast, the United Churches, mainly through the Baptists and Churches of Christ, who were then training more ministers than they could employ,[70] had no shortage of younger clergy interested in military chaplaincy. Through the 1990s it was able to fill vacancies that the larger denominations could not.[71] It was a trend that was already

---

67 Michael Gladwin, *Captains of the Soul. A History of Australian Army Chaplains* (Sydney: Big Sky, 2013), 265-269.
68 RACS, *Minutes,* 13 December 1989; FUCCB, *Minutes,* 7 August 1990.
69 RACS, *Minutes,* 19 June 1990; FUCCB, *Minutes,* 7 August 1990.
70 Dennis Nutt, *A Crucible of Faith and Learning: A History of the Australian College of Ministries* (Sydney: ACOM, 2017). Chapter 8 describes the expansion of ministry training in Churches of Christ during the 1990s.
71 Gladwin, *Captains of the Soul,* 271.

evident in 1991 when the total number of UC chaplains in the ADF had risen from thirty to forty-five over the preceding eight years.[72] By the end of the century, when census figures showed the United Churches as having about the same number of adherents as the Presbyterians and half that of the Uniting Church,[73] there were almost twice as many UC chaplains serving in the Reserve forces as Uniting Church chaplains, and three times as many as the Presbyterians.[74]

The Changing of the Guard

The last two decades of the twentieth century saw the emergence of the United Churches as a significant contributor to leadership of military chaplaincy. Whereas Allen Brooke and Walter Albiston had made distinguished contributions to chaplaincy administration, their positions had been primarily related to denominational seniority. The new professionally led Defence Force chaplaincy meant that administrative responsibility was now trans-denominational. McCullough, Crossman, Griffiths, Jarvis and Sabel all played significant parts in this.

McCullough was the link with what had gone before. In 1981, with the disbanding of the Conference of Chaplains General, he became the UC representative on RACS. Prior to that he had had a long and distinguished chaplaincy career that began in 1939 and culminated in his appointment as Chaplain General in 1964. FUCCB paid tribute to his forty-two years of Army service, referring to his graciousness, ready sense of humour and loyalty to both his Church and the Service:

> Dedication was the hallmark in his long, continuous period of service. His ministry to the Service has been distinguished by his pastoral, caring concern for both his chaplains, and to the many soldiers, to whom he was called to minister.[75]

---

72 FUCCB, *Minutes*, 13 October 1981; 23 March 1982; 30 March 1984; 29 June 1992.
73 Stephen Reid, 'Australia's Religious Communities; Numbers of people identifying with selected religious groups, 1911–2016, *Christian Research Association*.
74 FUCCB, *Minutes*, 28 April 1999.
75 FUCCB, *Minutes*, 26 May 1981.

He was succeeded as UC representative on RACS in 1985 by Geoff Crossman, who had recently retired as Principal Air Chaplain – the first member of the Air Force to serve on this body.[76] Along with Griffiths, Jarvis and Sabel, he had represented the United Churches at the highest administrative levels of Australian military chaplaincy through the ADF's critical first two decades.

Griffiths' appointment in 1981 as the first Army Principal Chaplain (Protestant Denominations) was an historic event because it was the first time that all Protestant chaplains came under the denominational authority of one Protestant Principal Chaplain. His appointment, which required the support of all members of RACS, is an acknowledgement of the high esteem in which he was held, despite his chaplaincy experience having been part-time. As a member of the Principal Chaplains Committees, he helped guide ADF chaplaincy through this period of transition. Later, as FUCCB chairman and representative on RACS, he opened the way for the continuing expansion of United Churches chaplaincy in all three services, sensitively negotiating with other members RACS to fill positions they were unable to do.

He was still serving in this role when in 1987 Jarvis was appointed Principal Chaplain (Protestant Denominations) for the Navy, a position he held until his retirement from chaplaincy in February 1991. Then, on 1 February 1988, contrary to usual conventions of denominational rotation, Sabel replaced Griffiths as Protestant Principal Chaplain for Army. This was unexpected because the usual practice of denominational rotation should have meant that Bruce Roy (Uniting Church) would have been the most likely person to be appointed.[77] RACS, however, decided not to nominate Roy because of his medical classification.[78] This seems strange because it didn't preclude Roy, who had previously served in Vietnam, from continuing service as Command Chaplain for the Army's Training Command. Roy himself believed that the real reason was that his status as a divorced person was

---

76 FUCCB, *Minutes*, 22 October 1985.
77 RACS, *Minutes*, 3-4 September 1986.
78 Salvation Army Australian Southern Territory Headquarters, letter to Federal United Churches Chaplaincy Board, 29 March 1987, Baptist Union of Victoria Archives.

unacceptable to some members of RACS.[79] Consequently, Sabel was nominated and became the primary administrator of the Army Chaplains Department.[80]

Unfortunately, Sabel's role in the previously mentioned restructure of Army chaplaincy, which was enthusiastically received by most chaplains, appears not to have received the same level of acceptance by RACS. Sabel, having obtained the approval of the Army's Chief of Personnel to proceed with the restructure, informed RACS of what he intended and sought its comments.[81] That he did not seek its approval appears to have activated a controversy that had been simmering since 1981.

This writer's memory of the events, based on personal conversations with both Crossman and fellow chaplains, as well as later talks with Sabel, recalls that the decision for a restructure activated a demarcation dispute between RACS and the Principal Chaplains. It was an issue that had been lying dormant for a decade. Nothing of what followed was recorded in any minutes, apart from a reference to the suspension of the Standing Orders, which required the Principal Chaplains to leave the meeting while the members of RACS met in private,[82]

The Religious Advisory Committee had only recently transitioned from being a Conference of Chaplains General, whose members held high rank in the Army and were responsible for the administration of Army Chaplaincy. Now it was a body of civilian church representatives appointed by the Minister of Defence. Whereas the Chaplains General had been responsible for the internal management of Army chaplaincy, that role in all three services now belonged to the Principal Chaplains. Consequently, it appears that the restructure of Army chaplaincy, soon followed by that of Navy and Air Force, brought things to a head. The Religious Advisory Committee, or at least some members of it, felt that things were being done without their permission. They gave Crossman who, as the United Churches representative, was Sabel's bishop equivalent, the task of meeting the Army's Chief of Personnel and asking him to withdraw Sabel's appointment as Principal Chaplain.

---

79 Author's recollection of a conversation with Bruce Roy, 1987.
80 RACS, *Minutes*, 9 December 1987.
81 RACS, *Minutes*, 12 September 1989.
82 RACS, *Minutes*, 20 March 1990.

The Chief of Personnel, however, informed Crossman that it was not the role of RACS to interfere with what he, as Chief of Personnel, had approved. He also informed him that he had received many letters from chaplains supporting Sabel and threatening resignation if Sabel were to be removed.[83] Indeed, RACS itself also received such letters.[84]

Consequently, this very sensitive situation occasioned a special meeting between the members of RACS and the Principal Chaplains to deal with the tensions that had arisen. The outcome seems to have been a consensus that the members of RACS had 'ecclesiastical' oversight and control of chaplains, including nominating them, monitoring their service and facilitating their return to civilian duties,[85] while the Principal Chaplains had 'technical' control. It raised again the perennial question of who actually controls military chaplains: the military or the Church? The answer, of course, is both. However, in situations where there is a conflict between the two, the Church's only resort is to withdraw its chaplains, and in this case the chaplains themselves might possibly have rebelled in favour of the military.

It was unfortunate that at the end of a long and distinguished life of service circumstances caused Crossman to bear the brunt of this conflict. The general opinion among serving chaplains at the time was that he was the reluctant mouthpiece for other members of RACS. Crossman himself was known to have a high regard for Sabel, who bore him no continuing ill-will. Crossman retired from RACS on the last day of 1991 and was recognised by FUCCB with a special minute of appreciation, which described him as: 'A sound administrator ... A genial conversationalist ... A Christian gentleman [who] served his Lord and our Churches with distinction'.[86] The minute also mentioned that in 1982 Crossman had been made a Commander of the Order of the British Empire which, along with that awarded to Allen Brooke, was the highest award ever made to a UC chaplain.

---

83 Ernest Sabel, email to Robert Smith, 7 August 2021.
84 North Queensland Area Chaplains' Committee, letter to RACS, 9 April 1990, Baptist Union of Victoria Archives.
85 RACS, *Minutes*, 10 September 1990.
86 FUCCB, *Minutes*, 2 July 1991.

Sabel and Jarvis also announced their retirement from full-time chaplaincy that year, both transferring to the reserves and serving on as Reserve Principal Chaplains. [87] The departure of these three outstanding leaders marked the end of an era. Just as McCullough had been the bridge between those who served in World War II and the generation that followed, Crossman, Sabel and Jarvis were the bridge linking the pain of Vietnam with the pride of a new Australian Defence Force.[88]

For Australia's armed services, the period between the Vietnam War and the new century was a time of transition, both organisationally and in the level of esteem in which they were held. It was also a time of significant transition for Australian military chaplaincy, and the United Churches had played a major part in it. In 1972 the United Churches had still been the junior member of the joint Protestant contribution to chaplaincy. But by the end of the century, it was on its way to becoming the largest contributor, doubling its number of full-time Army chaplains, contributing significantly to chaplaincy leadership and finally making its presence felt in the Navy. It was now ready to take its place as the emerging cornerstone of Australian military chaplaincy.

---

87   RACS, *Minutes,* 10 September 1990; FUCCB, *Minutes*, 9 April 1991.
88   Sabel was made a member of the Order of Australia.

CHAPTER NINE

The War on Terror: The Cornerstone

THE START OF A NEW CENTURY saw the Australian Defence Force at its highest operational tempo since the Vietnam War. The wilderness years described in the previous chapter had faded by 1989 when a series of effective peace-keeping missions to Namibia, Cambodia, Somalia and Rwanda restored its reputation. The sixteen thousand strong multinational peace-keeping force to Cambodia (1991–1993), led by Lieutenant General John Sanderson, included more than five hundred Australian troops. It was the largest command by an Australian general since World War II.[1] Smaller contingents were also sent to the Middle East, Bougainville and the Solomon Islands, the latter lasting from 2000 to 2017.[2]

It was the Australian-led deployment of the International Force for East Timor (INTERFET) in 1999 that provided the biggest challenge. More than five thousand Australians took part in this operation. They quickly secured the capital, Dili, before moving out to other areas, disarming militia and sometimes having tense stand-offs with the Indonesian military. With calm restored, a large United Nations

---

1 Jeffrey Grey, *A Military History of Australia* (Melbourne: CUP, 2008), 269.
2 Grey, *A Military History,* 267-68.

peacekeeping force, including Australian troops, took over in February 2000 and remained there until 2013.[3]

INTERFET more than anything else redeemed the image of the Australian Defence Force in public esteem following the debacle of Vietnam, and the Anzac legend returned to its honoured place in the national psyche. But it also placed an enormous demand on the military in that it coincided with the ADF's support for the Sydney Olympic Games held in 2000. Joint Task Force Gold, whose headquarters was established at Victoria Barracks, Sydney, early in 1999,[4] included around four thousand personnel drawn from each of the three services and provided specific security and support roles. As part of his role at the Land Headquarters, Robert Smith (Churches of Christ) was posted to it as Senior Chaplain.[5] Admiral Chris Barrie, the Chief of Defence, remarked on the pressure placed on the Defence Force by having to mount two major operations simultaneously:

> This time last year, we contributed over 4500 personnel on operations in East Timor and now we are contributing over 4000 ADF and civilian personnel to Operation Gold ... Our people are performing a variety of roles at the games and are working magnificently ... And we are maintaining about 1500 in East Timor and personnel on other operations overseas.[6]

The pressure continued for the next two decades. Terrorist attacks that destroyed New York's Twin Towers on 11 September 2001 launched the American-led War on Terror which, with Australian support, led to both nations' longest running wars. The initial focus was on Afghanistan, but from 2003 to 2009 it included the war in Iraq. More than forty thousand Australians served in the Middle East Area of Operations.[7] Thousands more were involved in deployments to East Timor and the Solomon Islands. In these operations fifty-six Australians died between 1999 and 2017: forty-six in Afghanistan, five in East Timor, four in Iraq and two

---

3   Grey, *A Military History*, 276-278..
4   *Operation Gold, ADF Support to the Sydney 2000 Games* (Canberra: Australian Defence Force, 2000), 18.
5   Australian Defence Force, *Operation Gold*, 135.
6   *Army*, 28 September 2000, 4.
7   'Official Histories – Iraq, Afghanistan & East Timor,' Australian War Memorial.

in the Solomon Islands. Lieutenant General Peter Leahy later described the formal departure of Australian troops from Afghanistan as 'an enormously significant occasion' signifying 'the end of an era for the ADF'.[8] It was an era that stretched the Australian Defence Force to its limits.

Chaplaincy Support to Operations

Of the two hundred and twelve chaplaincy deployments between 2000 and 2017, sixty-one (29%) involved chaplains from United Churches.[9] Some had multiple deployments, including Mark Willis (Churches of Christ),[10] who served on six operations between July 2003 and January 2016, making him probably the most operationally experienced chaplain since World War II.[11] The only casualty among them was Gary Doecke (Lutheran), an Air Force chaplain who suffered a massive heart attack in Iraq and was repatriated to Perth. He survived on life support in Perth for more than eighteen months until he died in December 2006.[12]

Russell Mutzelburg (Churches of Christ)[13] was the first UC chaplain to be deployed since Vietnam. He served on Operation Bel-Isi, a peace monitoring operation on Bougainville in 1997.[14] He was followed by Craig Willmott (Churches of Christ) and David Grulke (Lutheran) from Army, Barrie Yesberg (Churches of Christ) from Navy, and Ian Whitley (Baptist) from Air Force. Willmott and Grulke were deployed to East Timor, Yesberg to the Persian Gulf and Whitley to Bali (following the Bali terrorist bombing) and the Middle East.[15]

Willmott is of particular significance in that he was the first Army Reserve chaplain to be deployed on active service since the creation of the Australian Regular Army in 1947.[16] In addition to his service

---

8   'Final Australian Troops Leave Afghanistan as 20-year Mission Draws to a Close,' ABC News..
9   Jason Wright, email to Robert Smith, 20 March 2023.
10 Later appointed Director General Chaplaincy – Air Force.
11 Mark Willis, email to Robert Smith, 9 February 2023.
12 FUCCB, *Minutes,* 20 May 2005 and 8 June 2007.
13 Later appointed Principal Chaplain – Army.
14 Russell Mutzelburg, email to Robert Smith, 8 February 2023.
15 FUCCB, *Minutes,* 26 October 2002.
16 Except when individual reservists were put on to *Continuous Fulltime Service*, the Army Reserve (before 1980 known as the Citizens Military Force) were not deployed overseas until 2007 when it assumed responsibility for Operation Anode in the Solomon Islands.

in East Timor, Willmott also deployed to Iraq in 2006 as part of the Al Mathana Task Group. He achieved another first in 2003 when he was posted with an Army detachment as chaplain on HMAS *Kanimbla*, becoming the first Australian Army chaplain to become part of a Navy ship's company.[17] He was followed by several other Reserve chaplains from United Churches who served with more than two thousand Reservists deployed to the Solomon Islands,[18] as well as some Air Force Reserve chaplains who went to the Middle East.[19]

Another significant first for United Churches was the appointment of Ivan Grant (Baptist) as the ADF's first indigenous chaplain.[20] Grant, who was deployed with the Army to Iraq in 2004,[21] came from a family with a proud military history stretching back to his great uncle, Ivan Grant, who being under-age enlisted under his older brother's name and was killed in action in France in 1917. His two grandfathers, Cecil Grant, a Wiradjuri man, and Noel Nichols, who was of European extraction, both served in the Australian 9th Division and fought together at the Battle of El Alamein. Grant says of them:

> They had different backgrounds, but they shared a common experience of war ... What they didn't know then was that [they] would cross paths again and become good friends ... In the crucible of war the colour of your skin suddenly becomes less important. What counts is that you can trust the person standing next to you with your life.[22]

In 1967 with the Vietnam War raging, and conscription again dividing the community, Grant's father Herb volunteered to be part of the ballot that determined who would be conscripted. As an Aboriginal man he was exempt from the ballot because Aboriginal people were not then recognised as citizens. Yet, following family tradition, he volunteered

---

17 Craig Willmott, email to Robert Smith, 9 October 2022.
18 *Solomon Times,* 31 October 2007, 1; 'Australia-led Combined Task Force concludes role with RAMSI,' Department of Defence Media Release, 2 July 2013.
19 Privacy restrictions prevented access to details of names but are known to include Fred Davis, Robert Smith, Garry Towle, Haydn Parsons and Mark Butler (Churches of Christ).
20 FUCCB, *Minutes,* 26 April 2002.
21 FUCCB, *Minutes,* 13 May 2004.
22 Ivan Grant, 'My Country, My Story – The Journey of an Aboriginal Army Chaplain,' *AACJ,* 27 (2016) 103.

to go to war for the country that denied him citizenship and fought in Vietnam.[23] These family traditions, which Grant says were 'shaped ... most powerfully by the Christian faith', eventually led to Ivan's ordination and later commissioning as a chaplain. In addition to his important pastoral ministry to indigenous soldiers, he has had an important role in providing cultural education to serving personnel and advice to ADF leaders, who see him as an important resource in matters relating to indigenous members.[24]

The United Churches also played a part in the emergence of the ADF's first female principal chaplain. Following the appointment of its first female chaplain, Wendy Snook (Uniting Church), to the Air Force in 1990,[25] Catie Inches-Ogden (Churches of Christ) was one of the first female chaplains to be appointed to the Army Reserve. In 2003 she transferred to the Regular Army,[26] becoming the first full-time female chaplain and served in a number of significant postings including a deployment to the Middle East: one of the first female chaplains to serve on operations.[27] She later transferred from the United Churches to the Anglican Church and was ordained as a priest. She continued her military chaplaincy full-time and advanced to Senior Chaplain and Command Chaplain roles, before finally being appointed Anglican Principal Chaplain.

The next female chaplain appointed from United Churches was Mairi Mitchell (Salvation Army) who joined the Army Reserve in 2007.[28] She was followed in 2011 by Melissa Baker (Baptist), who became the Navy's second female chaplain.[29] Her story of how she went from being a homeless teenager to an ordained minister with an earned doctorate is told in her books: *A Two-Way Street, Painting Beauty with the Ashes,* and *Conflict to Hope.* Sadly, her service in the Navy was cut short by health problems exacerbated by naval service.

---

23 He volunteered prior to the May 1977 referendum that gave Aboriginal people full citizenship.
24 Grant, 'My Country, My Story,' 105.
25 RACS, *Minutes,* 10 September 1990.
26 FUCCB, *Minutes,* 19 December 2003.
27 Kay Ronalds (Uniting Church) was the first Australian female chaplain to operationally deploy, serving in the Solomon Islands in 2005.
28 FUCCB, *Minutes,* 22 November 2007.
29 FUCCB, *Minutes,* 10 November 2011.

Despite these initial successes, further progress in recruiting female chaplains was disappointing. In 2017, of the ninety-three UC chaplains in the ADF only four were female. However, several others were in various stages of the recruiting process, raising hopes for an increasing female participation[30] in accordance with the ADF's desire for the number of female chaplains to be proportionate to its total number of female members.[31]

Moral and Spiritual Injury: a New Challenge for Chaplaincy

The wars in Afghanistan and Iraq provoked renewed concerns about issues related to the morality of war and warfighting. Reports of prisoner abuse and civilian casualties – euphemistically referred to as 'collateral damage' – began to emerge and continued throughout the duration. The most infamous was the abuse and torture of Iraqi prisoners of war held in Baghdad's Abu Ghraib prison. An Associated Press report in November 2003 led to an internal investigation that finally became a public scandal when a U.S. television news program aired a segment in April 2004 that included several photographs of prisoners undergoing abuse. It was followed by other disturbing reports, including an International Criminal Court finding, that 'hundreds of Iraqi detainees were abused by British soldiers between 2003 and 2009'. Then, in 2016, allegations of war crimes by Australian soldiers led to the Brereton Report which uncovered credible information that members of the Australian Special Forces had committed war crimes in Afghanistan between 2005 and 2016.

These revelations sent shock waves through the governments and military establishments involved and led to a renewed emphasis in the ADF on the moral dimension of ethical behaviour in military operations. Prior to this, in 2010, following the Abu Ghraib scandal, on a recommendation from the United Churches member,[32] RACS agreed to commission a paper on the ethical issues surrounding the use of

---

30 FUCCB, *Minutes*, 12 September 2017.
31 FUCCB, *Minutes*, 10 March 2017.
32 Author's reminiscences.

torture, as a resource for chaplains.[33] Fr Gerry Gleeson, a specialist in moral theology,[34] subsequently produced a paper defining torture and the appropriate response to it. This paper was then sent to the ADF Chaplains College to be incorporated into its unit dealing with the chaplain's role as a provider of ethical advice to commanders. It was also sent to the Centre for Defence Leadership and Ethics with a recommendation that it be developed into a course for ADF officers.

Furthermore, RACS expressed concern about the ability of existing programs for managing psychological stress to address the deeper spiritual issues associated with guilt and moral trauma. Consequently, it agreed that the chaplain posted to the Mental Health, Psychology and Rehabilitation Branch should be tasked with the development of a model of spiritual health, and that this should become the primary emphasis of chaplains posted to that position.[35]

This task was eventually given to Barrie Yesberg (Churches of Christ) who had been posted into a new position at Headquarters Joint Health.[36] The result was a strategy entitled *Towards Spiritual Health: The Australian Defence Force Spiritual Health and Wellbeing Strategy*. Its strategic objectives were: 1) To promote and support spiritual health and wellbeing, fitness and resilience in the Australian Defence Force; 2) To identify and respond to the spiritual health and wellbeing risks of military service; 3) To deliver comprehensive, collaborative, and appropriate spiritual health care; 4) To continuously improve the quality of spiritual health and wellbeing delivery; 5) To build an evidence base for spiritual health and wellbeing; and 6) To develop strategic and civilian partnerships regarding spiritual health and wellbeing. The strategy was presented to RACS in September 2014 and formally launched by the Chief of Defence soon after.[37]

Another factor that influenced this emphasis on spiritual resilience

---

33 FUCCB, *Minutes*, 6 May 2010.
34 Gleeson was also an Army Reserve chaplain.
35 FUCCB, *Minutes*, 18 November 2010; Assistant Minister for Defence, letter to Robert Smith, 10 December 2014; Gerald Gleeson, 'A Christian Response to Torture,' unpublished paper, 2010.
36 FUCCB, *Minutes*, 19 April 2012.
37 Barrie Yesberg, 'Towards Spiritual Health: The Australian Defence Force Spiritual Health and Wellbeing Strategy,' unpublished paper, 15 September 2014.

was the International Chiefs of Chaplains Conference held in Prague in February 2011. The four Australian delegates to the Conference included two members from United Churches: Robert Smith (Churches of Christ), who was by then the United Churches representative on RACS, and Garry Lock (Baptist), the Director General of Chaplaincy for Navy.[38] The theme of the conference was *The Military Chaplain as Ethical Advisor.* Its subject matter was a response to the revelations of abuse already mentioned. In a report to RACS, Smith referred to the Conference's strong emphasis on the ethical advisory role of chaplains, and their importance in responding to the 'moral injury' sustained by service people who have been involved in incidents that have violated their sense of basic values. The prophetic role of chaplains was highlighted in the need for them to oppose the extreme pragmatism that often allows abuse to take place in order to achieve short-term benefits, without considering the long-term ramifications. The conference also stressed the need to equip chaplains to convey such concepts effectively to commanders, and to prepare chaplains spiritually to survive in environments of moral trauma.[39] These initiatives contributed to a growing awareness within Defence of the vital importance of ethical conduct in military operations and the importance of developing spiritual resilience in its personnel.

Demise of the Proportionate Representation Policy

Chapter Two described how the 1912–1913 meetings that formalised chaplaincy arrangements in the Navy and Army and developed agreements about chaplaincy numbers were in 1914 amended to a system of proportionate representation based on religious affiliation. It was, as Michael Gladwin reflects, 'an imperfect system... [but had] few more equitable alternatives'.[40] It was, however, a flawed system because census figures did not represent the true nature of Australian religiosity. Tom

---

38 FUCCB, *Minutes,* 11 April 2011.
39 Robert Smith, 'Report on the International Military Chiefs of Chaplains Conference – Prague 31 January to 4 February 2011,' *RACS, Minutes,* March 2011.
40 Michael Gladwin, *Captains of the Soul. A History of Australian Army Chaplains* (Sydney: Big Sky, 2013), 33.

Frame, referring to the shortcomings of census figures in the first half of the twentieth century as an accurate indicator of religious commitment, concludes: 'Plainly, many people ... identified with a denomination merely for the sake of completing the [census] form'.[41] Moreover, at certain critical times[42] the policy was unable to recruit the required number of chaplains to meet the needs of Australia's armed services. Furthermore, the nature of operational environments required, and *de facto* received, chaplaincy support that was mostly trans-denominational and generic, rather than denominational.

The major churches have always been heavily overrepresented in census figures by nominal adherents who have little, if any, realistic commitment to those bodies. Stuart Piggin and Robert Linder make a number of references to this. Commenting on the 1947 census, they pointedly speak of 'large numbers of Australians ... [being] aware of the church or denomination they stayed away from', and that by the 1960s these had become more honest about this and 'defaulted to the secular, not to nominal Christianity as they had a century earlier'.[43]

The corresponding level of nominalism in the United Churches, however, has always been significantly less, as the National Church Life Surveys have demonstrated:

> Attendance rates in larger denominations such as Anglican (5%), Uniting (10%) and Catholic (11%) are a small proportion of the overall number identifying. By comparison, Other Protestant Denominations such as the Baptist Church (33%), tend to have much higher proportions attending. Pentecostal attendance estimates actually exceed Pentecostal affiliation figures from the National Census (102%).[44]

---

41 Tom Frame, *Losing My Religion – Unbelief in Australia* (Sydney: UNSWP, 2009), 89.
42 The two world wars and the current period since the 1990s.
43 Stuart Piggin and Robert Linder, *Attending to the National Soul* (Melbourne: Monash University Publishing, 2020), 23, 273.
44 Ruth Powell, Sam Sterland and Miriam Pepper, *The Resilient Church: affiliation, attendance and size in Australia* (Sydney: NCLS Research, 2020), 18. Pentecostal attendance figures exceeding actual affiliation figures are probably due to the high numbers of non-members attracted to Pentecostal services.

An even greater problem is the diminishing recruitment base that began to emerge in the late 1980s. The most obvious reason for this is the serious numerical decline of the major churches during the latter part of the twentieth century.[45] The Anglicans, who until the 1970s had the largest group of religious adherents, have over the course of a century declined by two thirds (39.7% in 1901 to 13.3% in 2016). Similarly, the Uniting Church between 1981 and 2016 dropped from 8.2% to 3.7%. The member denominations of the United Churches also experienced losses, but they were relatively insignificant, and the Baptists actually grew from 1.3% to 1.5%. On the other hand, the Pentecostals more than doubled from 0.5% to 1.1% and was 'the only group that has seen a consistent increase in affiliation'.[46]

Of even greater concern for the Roman Catholic Church was the seriously diminishing number of priests young enough and fit enough to be considered suitable applicants for military chaplaincy. In 2011, out of a national total of one thousand four hundred and forty-eight priests, only one hundred were under the age of thirty-five, and three hundred and fourteen under the age of forty-four. By comparison, the Baptist Church nationwide had three hundred and fifty-two ministers under the age of thirty-five and seven hundred and ninety-eight under the age of forty-four. Even more revealing is the figure for the Pentecostal churches, with five hundred and eighty-two ministers under thirty-five and one thousand one hundred and fifty-seven under forty-four.[47] Despite an agreement by RACS that the Roman Catholic Church be allowed to nominate trained lay pastoral associates as military chaplains,[48] and the service chiefs having given their assent,[49] there had been little change in the number of Roman Catholic chaplains. It is therefore not surprising that RACS was compelled to look to the United Churches to make up the deficiencies.

By the early years of the new century the situation had become

---

45 Except for the Roman Catholics, whose number of adherents remained steady because of large numbers of immigrants who declared themselves to be Roman Catholic.
46 Powell, Sterland and Pepper, *The Resilient Church*, 7-8.
47 Australian Bureau of Statistics, 'Census of Population and Housing 2011'.
48 It had already been given permission to nominate married deacons since the early 1990s; RACS, *Minutes*, 10 September 1990.
49 FUCCB, *Minutes*, 21 May 2009.

critical, prompting FUCCB in 2001 to agree that its representative on RACS should raise the subject of enlisting chaplains from the Assemblies of God, by then known as the Australian Christian Churches (ACC).[50] The motion he presented said:

> That FUCCB respectfully inform RACS that 1. We believe Chaplaincy provision is in crisis; 2. We believe that there is no ownership by the leadership of the respective churches of the need for chaplaincy services to the ADF; 3. We believe there is a need for RACS to involve all churches through their heads of Church in a broad ranging discussion of the issues of chaplaincy provision and the most appropriate structure to meet the new situation; 4. Once there is agreement amongst the Churches the Heads of Churches take up an urgent discussion with top level personnel in ADF re: the complete revision of the MOA.[51]

RACS, however, indicated that it was not prepared to proceed with the proposal, to which FUCCB recorded its disappointment.[52]

Nevertheless, the number of UC chaplains continued to increase as new candidates temporarily filled vacancies which the major churches were unable to fill. The outcome was that whereas in 1990 there were forty UC chaplains serving: eleven full-time, twenty-six part-time, and three on Continuous Full Time Service (CFT),[53] by 2008 the number had grown to sixty-two: thirty-one full-time, twenty-eight part-time and three CFT, continuing to grow to thirty-three serving full-time, and forty-two part-time in 2013.[54]

The result of this was that by 2013 all chaplaincy positions in all three services were filled, bringing praise to RACS from the official body whose task was to conduct regular reviews of such statutory bodies. It affirmed the importance of RACS in providing for the wellbeing of members of the ADF, particularly in having achieved 102 percent of

---

50   FUCCB, *Minutes*, 25 May 2001.
51   FUCCB, *Minutes*, 25 November 2001.
52   FUCCB, *Minutes*, 26 April 2002.
53   '1989, 1990 Denominational Statistics and Chaplain Numbers', in *United Churches Chaplaincy Historical Documents*, Baptist Union of Victoria Archives.
54   FUCCB, *Minutes*, 9 May 2013.

the number of chaplains required.⁵⁵ What the review did not reveal was that this result could not have been achieved under a strict observance of the proportionate representation system.

By the end of 2017 the figures were even more significant. The number of UC chaplains had by then increased to ninety-three: forty-seven full-time and forty-six part-time: an overall increase of 132 percent in twenty-seven years.⁵⁶ The United Churches had now become the largest provider of chaplains to the ADF, surpassing even the Anglican Church (traditionally the largest provider) which in 2017 had ninety chaplains serving, thirty-five full-time and fifty-five part-time.⁵⁷ It was now evident to RACS that the proportionate representation policy was dead. Service chiefs were making it clear that their concern was to have all vacant positions filled with competent chaplains, not with maintaining denominational balances.⁵⁸ By default, driven by the reality of a diminishing source of recruits from the major churches, Australian military chaplaincy had become generic rather than denominational. But there were four other factors that contributed to this fundamental shift in policy.

Protestant Ministry to the Australian Defence Force

In June 2007 the three Protestant members of RACS and the three Protestant Principal Chaplains formed an executive named *Protestant Ministry to the Australian Defence Force* (PMADF). It was to meet quarterly 'to give collective oversight to the interests of [Protestant] chaplains'. It also envisaged the possibility that the three part-time Protestant members of RACS might eventually merge into one full-time Protestant representative.⁵⁹ This would bring the Protestants into line with the Anglicans and Roman Catholics, whose representatives were full-time bishops located in Canberra, with closer access to ADF leadership than their scattered Protestant counterparts. Though this latter prospect failed to eventuate, the creation of the executive was a major

---

55   FUCCB, *Minutes*, 31 November 2013.
56   FUCCB, *Minutes*, 12 September 2017.
57   Defence Force Board Report to the General Synod of the Anglican Church, 2017.
58   Mark Willis, Director General Chaplaincy – Air Force, Interview with author, 16 February 2023.
59   FUCCB, *Minutes*, 8 June 2007.

step forward in ecumenical cooperation, building on the already established principle of Protestant cooperation for administrative purposes. In April 2011 the executive took a further step by deciding to include all Protestant applications for chaplaincy as one group when presenting them to RACS, rather than individually.[60] It proved to be a decisive moment in RAC's acceptance of what had for long been a practical reality: that the proportionate representation policy had ended.

The Emergence of the Australian Christian Churches

Mention has already been made of the significant growth of this denomination and its developing relationship with the United Churches, including the appointment of its first Army Reserve chaplain. The first decade of the twenty-first century saw this relationship grow significantly, even though David Griffiths (Baptist) had recommended that FUCCB should move slowly.[61] By 2007 all member denominations of FUCCB had agreed to support the inclusion of the Australian Christian Churches (ACC) in the Memorandum of Arrangements with Defence, and for it to have a representative on FUCCB. RACS subsequently agreed to recommend to the Minister for Defence and the Chief of Defence that ACC be invited to provide chaplains for the ADF.[62]

It is significant that one of Griffiths' last acts as FUCCB chairman was to welcome the ACC's first representative to the Federal Board. Griffiths, more than anyone, had worked towards this goal, recognising its importance for the future of Australian military chaplaincy. In a special minute, FUCCB acknowledged his fifty years of service and ministry in Defence and his long service as the Baptist Union of Australia's representative on the Board, including sixteen years as chairman. Robert Smith was appointed unanimously as his successor.[63]

The first ACC chaplain to be appointed (with the exception of Michael Alcock, who only served briefly) was Ralph Estherby. His was a

---

60 FUCCB, *Minutes*, 11 April 2011.
61 FUCCB, *Minutes*, 12 May 2006.
62 FUCCB, *Minutes*, 8 June 2007.
63 FUCCB. *Minutes*, 22 May 2008.

significant appointment, not only for his effective chaplaincy ministry[64] but also for his appointment in 2012 as denominational representative to FUCCB.[65] Estherby quickly showed himself to be an effective recruiter of new chaplains.[66] The first was Kees Bosch who joined a small but growing group of UC chaplains in the Navy.[67] By 2017 the number of ACC chaplains had increased to twelve: four in Army – two full-time and two part-time; three in Navy – two full-time and one part-time; and five in Air Force – four full-time and one part-time.[68] It was an indicator of things to come.

Multi-Faith Chaplaincy

The Royal Australian Army Chaplains Department, strictly speaking, has been a multi-faith body since 1943 when a Senior Jewish Chaplain was appointed (although a roving Jewish chaplain was commissioned during World War I). Australia's involvement in the Middle East, and concerns about its effects on the growing Muslim population within Australia, caused RACS to give serious consideration to ensuring that all faith groups within the ADF should have access to appropriate spiritual and pastoral support.

Its first step was to recommend to the Chief of Army the names of people who could act as Muslim advisors. The Chief of Army decided not to proceed with this at that time.[69] The scope was then widened to include Buddhists, following an expression of interest by the Buddhist Society in 'having input into religion within the ADF'.[70] Eventually, in 2010 RACS recommended the formation of an affiliated representatives committee, which would include representatives of other faiths with members serving in the ADF, and would be chaired by a member

---

64 He was awarded the Conspicuous Service Medal.
65 He was later to be appointed Chairman of the Federal Board.
66 Ralph Estherby, email to Robert Smith, 4 August 2022; FUCCB, *Minutes*, 18 November 2010 and 19 Apr 2012.
67 FUCCB, *Minutes*, 23 October 2012
68 FUCCB, *Minutes*, 12 September 2017.
69 FUCCB, *Minutes*, 8 June 2007.
70 FUCCB, *Minutes*, 11 December 2008.

of RACS.[71] It approached the Prime Minister's Department for names of suitable representatives, but received no reply. Consequently, despite the lack of response from government, it resolved to invite leaders of the Buddhist, Hindu, Muslim, Latter Day Saints and Sikh religions to join the proposed committee.[72] This proved successful, and the Affiliated Representatives Committee chaired by Rabbi Ralph Genende began to meet soon after.[73]

The process then became a little confused when the Chief of Defence asked RACS to support his desire for a Muslim representative to be appointed to it. The Committee reminded him that it had already created the Affiliated Representatives Committee for the very purpose of providing a voice for other faiths, including some that had larger numbers of members in the ADF than the Muslims. It did, however, agree to the appointment of a Muslim representative and noted that the required amendment to the MOA ought also to include the appointment of representatives for those other faiths with significant members in the ADF.[74]

A decision to appoint an imam became public in 2015[75] and was immediately reported by the national press. *The Age* reported: 'The ADF is set to get its first Muslim imam as part of a push to attract more recruits from different cultural and language backgrounds'.[76] The *Australian* announced: 'The Australian Defence Force has been told it must quickly take on a more culturally diverse range of recruits to improve its fighting ability'.[77] And the *Herald-Sun* noted that: 'Defence is failing to attract recruits from diverse ethnic and language backgrounds, and there are only ninety-six Muslims in the ranks'.[78] No decision was made as to how many chaplains might be appointed but this development was a further step away from the system that had characterised Australian military chaplaincy since 1912.

---

71 FUCCB, *Minutes,* 18 November 2010.
72 FUCCB, *Minutes,* 23 October 2012.
73 FUCCB, *Minutes,* 21 November 2013.
74 FUCCB, *Minutes,* 27 November 2014.
75 Chief of Defence, letter to Australian National Imams Council, 11 March 2015.
76 *The Age,* 3 March 2015, 6.
77 The *Australian,* 3 March 2015, 4.
78 The *Herald-Sun,* 3 March 2015, 5.

Secular Chaplains

By far the most contentious challenge to the chaplaincy *status quo* was the question of appointing *secular* chaplains to meet the needs of non-religious ADF personnel. Official statistics revealed an ever-increasing number of Australians declaring themselves as having no religion, and in the 2016 Census it was 'larger than any other religious group'.[79] Philip Hughes observed:

> Prior to the 1933 Census, the percentage identifying as Christian had been about 96 percent. In 1933 it dropped to about 86 percent and stayed in that vicinity through to 1971. Since 1971, identity with a Christian denomination has fallen gradually and may well be below 60 percent in the 2016 Census.[80]

The actual figure turned out to be 52.1 percent, and the number of people claiming 'no religion' increased from 22.3 percent in 2011, to 30.1 percent in 2016. These figures were even more marked in the ADF, which is made up overwhelmingly of people under the age of thirty. This became the basis of a challenge to the chaplaincy *status quo* by Colonel Phillip Hoglin, who noted that whereas two thirds of ADF personnel nominated Christianity as their religion in 2003, by 2017 that figure had reduced to just under half. Conversely, over the same period the number claiming to have no religion had risen from just under a third in 2003 to slightly more than half in 2017.[81]

Philip Hughes challenges the assumption that the decline in religious adherence is purely a result of the rise of secularisation, arguing that 'much of the space taken by the decline of religious practices and beliefs … has been taken over by a more spiritualistic individuality'. There are now more people who describe themselves as 'spiritual but not religious' than simply 'religious'. Based on the 2016 Census, he posits 'a rough picture of the Australian population' as being 20 percent traditional religious, 5 percent post-traditional religious, 25 percent post-traditional

---

79 Powell, Sterland and Pepper, *The Resilient Church*, 7.
80 Philip Hughes, *Charting the Faith of Australians: Thirty Years in the Christian Research Association* (Melbourne: Christian Research Association, 2016), 31.
81 Phillip Hoglin, 'Losing Our Religion: The ADF's Chaplaincy Dilemma'.

spiritual, and 50 percent neither religious nor spiritual.[82] It is a trend that David Tacey saw emerging twenty years earlier. He wrote:

> Our secular society [is] realising that it has been running on empty, and has to restore itself at a deep, primal source, a source which is beyond humanity and yet paradoxically at the very core of our experience. It is our recognition that we have outgrown the ideals and values of the early scientific era which viewed the individual as a sort of efficient machine.[83]

These statistics are a serious challenge to the current chaplaincy system, which is still coming to terms with the rapidly changing nature of Australian spirituality.

Robert Smith, having encountered the fact of humanist chaplains in some European military organisations at the 2011 Prague conference, raised the possibility with RACS that this could have implications for Australian military chaplaincy also, particularly in the light of increasing 'no religion' statistics.[84] It finally emerged in a paper written by Principal Chaplain Colin Acton (Anglican), the Director General of Chaplaincy for Navy. In 2017, Acton wrote:

> In the space of just one generation, the ADF has gone from an overwhelmingly Christian based organisation to one that is now secular and increasingly pluralistic. Personnel indicating 'no religion' on census and human resource data are the largest single group in the ADF and, within just a few more years, this group will be numerically larger than all religious groupings combined.

He concluded his paper with a recommendation that a 'non-affiliated chaplaincy endorsing body' be created in addition to the Religious Advisory Committee to the Services, 'which currently only represents the religious affiliations of 38 percent of Defence Force personnel'. His idea was that this additional body would source new chaplains from a

---

82 Philip Hughes, *Charting the Faith of Australians*, 32, 37.
83 David Tacey, *The Spirituality Revolution: The Emergence of Contemporary Spirituality* (Sydney: Harper Collins, 2004), 1.
84 Author's reminiscences.

wider range of suitably qualified people other than ordained persons, thereby ensuring more effective support for female, religious minority and non-religious Defence personnel.[85]

RACS, while agreeing with much of the intent of Acton's paper, argued that it was already dealing with many of the issues raised, particularly in relation to minority faith groups and attempts to recruit more female chaplains. It was not prepared to support the proposal for a non-affiliated chaplaincy endorsing body without having given it further detailed consideration.[86]

Nevertheless, the Navy was suitably impressed by Acton's arguments and decided to create a new position designated *Maritime Spiritual Wellbeing Officer,* whose background training and experience would be in social work, psychology, counselling, occupational therapy, and mental health nursing.

To date the other services have shown little interest in following the Navy's decision, believing that their current arrangements through chaplains and psychologists provide an appropriate range of support to personnel. Nevertheless, the Navy's action does indicate that the chaplaincy *status quo* is changing significantly. It can, of course, be argued that the concept of non-religious and minority-faith chaplains might indicate a return to sectionally-based chaplaincy in another form, rather than the emerging generic form. This, however, is not necessarily so. As far back as the early 2000s the Army's sole Jewish chaplain in Eastern Region, Yossi Segelman, was posted from a regional Jewish-specific role into a unit where he was required to provide chaplaincy support to all members.[87] The same could, and probably would apply to chaplains of other minority faiths, assuming that their training and experience in general pastoral care is equal to those required of existing chaplains.

Acton, in a later paper, noted that the role of chaplains had 'substantially changed from the provision of religious services ... to a more pastorally and spiritually focused caregiver'. He referred to Defence

---

85 Colin Acton, 'Increasing Diversity Within the Navy Chaplain Branch: Respecting the Past, Creating the Future,' unpublished paper, 15 November 2017, Baptist Union of Victoria, Archives.
86 RACS, 'Comment on DGCHAP Navy OUT/2017, Increasing Diversity Within the Navy Chaplain Branch,' undated, Baptist Union of Victoria, Archives.
87 Author's reminiscences.

Chaplain Pay cases in 2002 and 2013, which were largely predicated on 'responsibilities to provide Mental Health support, advanced pastoral care, spiritual support and moral/ethical advice'. He noted that data indicated that 'up to 95 percent of work undertaken by chaplains is pastoral care, mental health and spiritual support'.[88]

This being the case, apart from faith-group specific support, what chaplains do is largely generic in nature, and it has long been part of Australian military chaplaincy praxis for local chaplains to arrange for faith-specific support that they are unable to provide, to come from a chaplain or religious practitioner who can, when the occasion requires it. The same would be required of a non-religious chaplain and would, essentially, be little different from the current working relationship between ADF chaplains and psychologist officers. Nevertheless, apart from the Navy's Maritime Spiritual Wellbeing Officer there has, thus far, been no attempt to recruit a non-religious chaplain in any of the three services.

It is unlikely that the question of non-religious chaplains will go away. As the number of ADF personnel claiming to have 'no religion' continues to grow, and the gap between traditional Christian/religious values and those of mainstream Australian society widens, it seems inevitable that the traditional chaplain's role as guardian of and adviser on morality will be shared with others who espouse a different set of values. This is nowhere more obvious than in Australian society's growing acceptance of diversity in relation to sexuality. This may well pose the greatest challenge to Australian military chaplaincy's traditional role as subject matter expert in relation to morality.

The ADF is now committed to diversity, especially with respect to lesbian, gay, bisexual, transgender and intersex (LGBTI) personnel. This sits uneasily with most of the churches involved in chaplaincy and with many chaplains, who believe that such sexual orientations are contrary to the will of God. Though the ADF's traditional rejection of homosexuality ceased in the early 1990s, the issue for chaplaincy came to a head in 2005 when Defence Policy changed to accept same-sex

---

88 Colin Acton, 'Navy Chaplain Branch – Maritime Spiritual Wellbeing Officer (MSWO) Stream Position Paper, unpublished paper, 26 October 2018, Baptist Union of Victoria, Archives.

relationships equally with heterosexual relationships, and for them to be included in the relationship development courses that chaplains conducted. FUCCB expressed considerable concern over this decision and resolved:

> Because of the potential in this decision for chaplains to be put into situations which may compromise their own consciences and/or teaching of their churches, the FUCCB supports the recommendation that chaplains not be required to lead relationship development courses involving same-sex couples, and that such courses be given to the Defence Community Organization to conduct, leaving chaplains to conduct those courses that relate to couples living in traditional marriage relationships.[89]

The ADF accepted this recommendation, and chaplains were subsequently free to decide for themselves whether they would include same-sex couples in their courses. Nevertheless, this matter continued to dominate public debate with the growing demand for acceptance of same-sex marriages. It culminated in the national plebiscite that saw 61.6 percent of participants vote in favour of changing the marriage law to include same-sex unions, and the subsequent changing of that law by act of Parliament on 7 December 2017.

This exposed the ever-widening gap between traditional Christian teaching and that of contemporary society. Not all churches were opposed to the changes, neither were all ADF chaplains. But the majority of conservative and evangelical chaplains were, including many from United Churches. This may well prove to be the greatest challenge to that body's increasing significance in Australian military chaplaincy. It highlights the widening gap between what conservative Christians believe to be essential Christian moral values and those of contemporary society, which now considers them to be reactionary and discriminatory.

---

89 FUCCB, *Minutes*, 18 November 2005.

Growing Contribution to Chaplaincy Administration

Throughout this period, UC chaplains increasingly filled many senior chaplaincy positions, particularly in Army and Air Force. Russell Mutzelburg (Churches of Christ) was appointed full-time Principal Chaplain for Army[90] in January 2007 and served in that role until the end of 2010, when he transferred to the Army Reserve and became the part-time Protestant Principal chaplain. Also in 2007, Garry Lock (Baptist) was designated the next Director General of Chaplaincy for Navy, occupying that position until 7 February 2012, when he also took up the part-time Protestant Principal Chaplain appointment.[91] Mark Willis (Churches of Christ), following his posting as a Command Chaplain from 2013–2015, was appointed Director General of Chaplaincy for Air Force, and served in that role until January 2022. His brother, Peter Willis (Churches of Christ), was Command Chaplain for Joint Operations Command from 2014–2015[92] and Barrie Yesberg (Churches of Christ) became Protestant Principal Chaplain for Navy from 2016–2020. These appointments, along with a disproportionate number of UC chaplains serving at Senior Chaplain (Division 3) level, made this a significant period for United Churches' contribution to Australian military chaplaincy leadership.[93]

The Associated Protestant Churches Chaplaincy Board

Smith retired as Chairman of FUCCB at the end of 2014. The November Board meeting recorded a special minute of appreciation, thanking him for his 'outstanding service as a member and Chairman of the Federal United Churches Chaplaincy Board', citing particularly 'the quality of his leadership, reporting, representation on the Religious

---

90 Later to be designated Director General Chaplaincy – Army.
91 FUCCB, *Minutes,* 30 November 2006, 11 April 2011. He was awarded the Conspicuous Service Cross.
92 Mark Willis, interview with Robert Smith, 16 February 2023.
93 The growth of United Churches/Associated Protestant Churches leadership continued after 2017 with Darren Jaensch (Lutheran) serving a second term as Director General Chaplaincy – Army, followed by Kerry Larwill (Baptist) in 2023; and Willis being replaced as Director General Chaplaincy – Air Force by James Cox (Baptist) also in 2023.

Advisory Committee to the Services and pastoral care of chaplains'.[94] His replacement was Garry Lock.[95]

Two highly significant events occurred during Lock's term as Chairman. The first was the addition of the Seventh Day Adventist Church (SDA) as a member of the group. The second was the change of its name from the Federal United Churches Chaplaincy Board to the Associated Protestant Churches Chaplaincy Board. The decision for the name change was influenced primarily by the tendency for people to confuse the United Churches with the Uniting Church. Addition of another member denomination provided an opportune time to make the change. Inclusion of the Seventh Day Adventist Church (which had previously expressed interest in 1982 and 1987) was a further indication of the growing acceptance of diversity within chaplaincy and recognition of the changing nature of religion in Australia. Consequently, FUCCB, at its September 2017 meeting, formally resolved to change its name to the Associated Protestant Churches Chaplaincy Board and to pursue the matter of the inclusion of the Seventh Day Adventist Church.[96] The following months saw those tasks successfully accomplished and the first SDA representative was welcomed to the newly named Associated Protestant Churches Chaplaincy Board in March 2018.[97]

These two events marked what could be described as a high-water mark in the history of what began as the Other Protestant Denominations. No longer were the Baptist and Congregational Unions, the Churches of Christ and Salvation Army relegated to the sidelines, fighting for acceptance in a religious environment dominated by the major churches. By the end of 2017, broadened by the addition of the Lutheran Church of Australia, the Australian Christian Churches and the Seventh Day Adventist Church, the Associated Protestant Churches Chaplaincy Board had become the largest provider of chaplains for the Australian Defence Force and arguably its most significant contributor to chaplaincy leadership. The once rejected stone had finally become the cornerstone.

94 FUCCB, *Minutes*, 27 November 2014.
95 Lock served until the end of 2020, Associated Protestant Churches Chaplaincy Board, *Minutes of Meeting*, 27 September 2019.
96 FUCCB, *Minutes*, 10 March 2017, 12 September 2017.
97 APCCB, *Minutes*, 9 March 2018.

CONCLUSION

IN HIS HISTORY OF Australian Army chaplains, *Captains of the Soul*, Michael Gladwin observes that the establishment of the Australian Army Chaplains Department in 1913 was quite different from that of its British parent, the Royal Army Chaplains Department. Though its ethos and antecedents were British, the Australian Chaplains Department was created with a multi-denominational leadership that reflected Australia's rejection of an established church and provided a level playing field for a variety of faith traditions.[1] This followed the formation of the Royal Australian Navy's Chaplaincy Branch one year earlier with a similar multi-denominational structure, and was itself followed by a similar pattern for the Royal Australian Air Force in 1926.

The reality, however, despite its inter-denominational ideal, was that the 'level playing field' was more level for some than it was for others. For most of the twentieth century it belonged almost exclusively to the major churches, which dominated Australian military chaplaincy from its beginning. Their dominance seriously restricted the ability of smaller faith traditions to contribute to chaplaincy, and, in the case of the Navy, eliminated it until the 1960s.

This study has demonstrated how a small, disparate group of Christian denominations, initially known as the Other Protestant Denominations, struggled to gain acceptance on that supposedly 'level playing field' and, through successive eras of Australian military history, earned the respect of the major churches, eventually, as the Associated Protestant Churches, becoming the largest provider of chaplains to the

---

1   Michael Gladwin, *Captains of the Soul. A History of Australian Army Chaplains* (Sydney: Big Sky, 2013), 326,

Australian Defence Force and, arguably, the cornerstone of its chaplaincy organisation.

A little over a century earlier, the major obstacle to its participation was its numerical significance. The 1911 Commonwealth Census reported nearly 84 percent of Australians claimed affiliation with the major churches, whereas the corresponding figure for the Other Protestant Denominations was a mere 5.4 percent.[2] The entrenched sectarianism of the day, which Gladwin describes as virulent,[3] severely restricted the group's attempts to contribute to military chaplaincy, confining it to the margins of the military organisation. That same sectarianism enabled the major churches to consolidate their positions in the chaplaincy establishments and impose an unwanted denominational system upon them, despite the Navy's original desire for a trans-denominational form of chaplaincy.[4]

The eventual agreement with the Army did, however, provide a seed-like opportunity for the OPDs which, like the mustard seed that Jesus described as being the smallest of seeds yet grows to be the largest of garden plants,[5] over the next century grew to be the mainstay of Australian military chaplaincy. It gave Military District Commandants the authority to appoint clergymen of 'any recognised religious body' as chaplains on their establishments.

Several Baptist and Congregational ministers were already serving under the *ad hoc* system that preceded the establishment of the Australian Army Chaplains Department in 1913 and the introduction of compulsory military training in 1911 opened the door for others. The newly appointed Chaplains General appear not to have raised any objection to such appointments, nor to those for the Australian Imperial Force soon after, provided they remained strictly proportionate to census figures. But they did object to the petition in 1915 for the appointment of an OPD Chaplain General. Their response, as expressed in the Presbyterian Chaplain General's description of the group as 'an

---

2   Commonwealth of Australia, Census. 3 April 1911.
3   Gladwin, *Captains of the Soul*, 26.
4   Rowan Strong, *Chaplains in the Royal Australian Navy, 1912 to the Vietnam War* (Sydney: UNSW Press, 2012), 42.
5   Matthew: 13:32.

aggregate of quite non-coherent and non-corporate factors having no solidarity or unified administration and responsibility', clearly reveals their reservations, as did Archbishop Riley's public demeaning of Baptist and Churches of Christ chaplains and the Methodist Chaplain General's implicit refutation of Salvation Army officers as equivalents of ordained ministers.

Proportionate representation thus became the determining factor in who was authorised to play on this 'level playing field'. Appropriate though it may have seemed, individual warships, at best, only have room for one chaplain, despite the diverse religiosity of the ship's company. Similarly, the basic units of army formations: infantry battalions, artillery regiments and so on, only have establishment positions for one chaplain, whose role is to minister to all. This, and the growing awareness of the value of ecumenical cooperation in witnessing to a community whose denominational allegiances were largely nominal, made the proportionate representation system far from ideal.

Nevertheless, proportionate representation continued to dominate chaplaincy until the end of the century, when the inexorable march of secularism, multi-culturalism and the diminishing strength of the major churches sounded its death knell. Yet its fundamental weakness – the inability of the major churches to meet their recruiting targets in times of major mobilisation – had been evident since World War I. This was compounded in both world wars by bishops who, fearing for the health of parishes left without pastoral leadership, refused to allow clergy to enlist.

Under the proportionate representation system, the Other Protestant Denominations were entitled to five percent of chaplaincy positions in the 1st AIF. However, it eventually provided 6.5 percent of the total, in addition to many others who served at home in military camps and hospitals. Though not a huge disparity, it was an indicator of greater problems to come. During the inter-war period the proportionate system almost broke down completely and, once again, the OPDs filled the gaps left by the major churches. But it was the experience of World War II that finally proved its weakness. The massive build-up of troops from 1942 onwards severely tested the capacity of the major churches to meet their goals. Once again, the Other Protestant Denominations (renamed the United Board in 1920 and United Churches in 1942) made up the shortfall, providing almost twice as many Air Force chaplains and more than three times as many Army chaplains as required of it.

Capability to meet overall chaplaincy shortfalls was not the only reason for the OPD rise to prominence. Gladwin comments that of the four hundred and fourteen chaplains who served with the 1st AIF: 'The majority served with distinction ... in the crucible of modern industrial warfare with its horrific destructive capacity'.[6] This was manifestly true of the OPD team. Between them they received twice as many awards as was the average for chaplains. Furthermore, included in their number were some of the most respected World War I padres, including MacKenzie, the man whom Charles Bean rated as 'foremost' among the chaplains and whom McKernan describes as 'the most popular man in the AIF'.

It was this factor, probably more than anything, that dispelled whatever reticence the major churches had about them. By the war's end the Other Protestant Denominations was accepted as a valued, though junior, partner in Army chaplaincy, even though it still had no part in Navy chaplaincy. This continued through the inter-war years and flowered again during World War II, when it also became an integral part of Air Force chaplaincy.

It was during this period that the group finally achieved equal standing with the major churches at the highest levels of chaplaincy. A United Board/United Churches Chaplain General was appointed for the Army and a Staff Chaplain for the Air Force. Except for a continuing adherence to the policy of proportionate representation, the United Board/United Churches had finally become an equal player on the 'level playing field' of Australian Chaplaincy, except in the Navy.

The post-World War II years, which saw a return to peace-time establishments, also saw the birth of the Australian Regular Army and the beginning of the professionalisation of military leadership. The Air Force, aware of its need for greater specialist skills, also began to move towards being a professionally staffed service. The Navy, which by necessity had been led by professional sailors, maintained this process.

6   Gladwin, *Captains of the Soul*, 32.

Military chaplaincy soon began to follow this trend. The major impetus was the creation in 1981 of Principal Chaplains, representing the Anglican, Roman Catholic and Protestant denominations, as the chaplaincy administrators of each service. The original intention was for one to serve full-time as chair of each Principal Chaplains Committee on a rotational basis. However, by the twenty-first century the diminishing number of experienced chaplains from the major churches, and demands by service chiefs for appointments based on merit rather than denominational balance, resulted in a disproportionate number of UC chaplains becoming Principal Chaplains.

Notable among them were Geoff Crossman, the first full-time Principal Chaplain for Air Force; David Griffith, the first Protestant Principal Chaplain for Army; Ern Sabel, who initiated the reorganisation of Army chaplaincy for a new age, and Ken Jarvis, the first full-time Navy chaplain and Principal Chaplain to be appointed from United Churches. The first two decades of the twenty-first century saw another disproportionate number of UC Principal Chaplains appointed: Russell Mutzelburg, Garry Lock, Barrie Yesberg, Darren Jaensch, and Mark Willis.[7]

Thus, the study has shown how this small and disparate group of churches not only fought for the right to participate in military chaplaincy but proved itself equal to the best the major churches provided. Furthermore, from its earliest days it helped develop an ethos that continues to undergird military chaplaincy: incarnational ministry, servant leadership, continuity of service, ardent evangelicalism, and practical ecumenism. Its contribution to the latter two has been particularly significant.

The study has traced its tortuous progress through the various eras of Australian military history and demonstrated how it has consistently met, and in times of greatest need surpassed all that was expected of it, both quantitatively and qualitatively. Between 1914 and 1946 it compensated for shortfalls in recruiting by the major churches, and from 1942 began to contribute increasingly to the senior leadership of chaplaincy. By the turn of the century, its contribution to both chaplaincy numbers and chaplaincy leadership had proved critical. Without

---

7   Jaensch was followed in 2022 by Kerry Larwill (Baptist), and Willis by James Cox (Baptist).

it, the Religious Advisory Committee to the Services would not have received officialdom's accolade affirming its importance in the wellbeing of members of the ADF particularly in having exceeded the number of chaplains required.

The study has also highlighted the group's importance in relation to the latest challenge to the chaplaincy *status quo*. The National Census of 2016 revealed that: 'For the first time in history, the proportion of Australians who had "no religion" was larger than any other religious group'.[8] This is even more evident in the ADF, which is made up overwhelmingly of people under the age of 35. It raises the question of whether chaplains from mainstream denominations, whose religious beliefs and world view differ from those of minority religion or no-religion groups, are now able to provide effective pastoral, ethical and spiritual advice, as is their role. Yet, it can be argued that chaplains have always represented a different worldview. 'In the system but not of it' has been chaplaincy's unofficial motto from the beginning, underlining its moral and spiritual independence from military authority even while living and working under it. Nevertheless, in an increasingly non-religious and multi-faith world, chaplains are under increasing pressure to prove their value to the Defence Force.

Despite decreasing levels of religiosity, the ADF continues to maintain its chaplaincy because generations of experience, including the most recent conflicts, have demonstrated its value in providing a dimension of support that other helping professionals may not access. One unique element in this is *hope*: existential hope in its transcendent dimension, as was expressed in a poster circulated widely through the Army some years ago with the caption: 'Army Chaplains – A Force for Hope'. A most significant contribution to this has been the Spiritual Health and Wellbeing Strategy developed by Barrie Yesberg, officially launched by the Chief of Defence in 2014.

Its relevance is seen in an incident reported by Morgan Batt (Roman Catholic), chaplain to an infantry battalion about to deploy to Iraq. During its pre-deployment training, Batt visited a temporary morgue containing several soldiers who had been designated as having been

---

8   Powell, Sterland and Pepper, The Resilient Church, 7.

killed in action. He checked their identity tags and gave the last rites to a Catholic, said prayers for a couple of Protestants and then found two whose tags said *no religion*. So, in sensitivity to their wishes, he did nothing. As he left the tent they sat up and said: 'What about us?' He replied that he was being respectful and not forcing unwanted religious ministration on them. Their reply was: 'Oh, we didn't know it meant that'. It was only later that Batt discovered the story had passed around the battalion and the quartermaster reported a steady stream of soldiers getting their identity tags changed from 'no religion' to one of the religious alternatives. They wanted to know that the padre would be there to pray for them, too. Existential hope, it seems, is still a concern, even for those with no religion.[9]

Today's military chaplains work in close cooperation with other professions providing a holistic system of physical, emotional and spiritual support. Good chaplains are like the wise elders of the tribe who are always there for people when they need to talk to someone. They are the face of humanity in situations where humanity seems to have died; a reminder that war and disaster are not the norm but that there is another and greater reality.

The factors that make for effective chaplaincy include intelligence, spirituality, training, theological competence, flexibility of personality, wisdom, the capacity to be a good listener, and much more. Though hard to define, effective chaplaincy is obvious when manifested in individual representatives. This study has demonstrated that from its beginning the Other Protestant Denominations (and its later incarnations) have been blessed with many.

Military historian David Dexter captures the ordinary soldier's assessment of a good chaplain in a reference to a soldier in the New Guinea campaign who described his Salvation Army padre as 'a reincarnation of the Biblical "Good Samaritan" ... [who] came not to reproach us for past sins or preach of the men we might have been ... [but] succoured the wounded and sick, and revived the tired and weary'.[10] Dexter also refers to a diary entry of the exhausted 2/32 Battalion, marching

---

9   Author's reminiscences.
10 David Dexter, *Australia in the War of 1939–1945; The New Guinea Offensives* (Canberra: Australian War Memorial, 1961), 489.

along the coast through drenching tropical rain to re-join its brigade, being met at the Burep Crossing by another Salvation Army padre with hot tea and biscuits. The entry records: 'To this noble institution the battalion once again tenders its thanks for another demonstration of practical Christianity'.[11] Chaplains like these transcend issues of religious belief or non-belief and speak in a language everyone understands. The story of the Other Protestant Denominations, as this study has shown, is full of them.

Even so, ADF chaplaincy is now in a state of flux and is likely to remain that way for the foreseeable future. The social realities of twenty-first century Australia have brought an end to the policy of proportionate representation, which for a century ensured the dominance of the major churches. The desire by Defence chiefs for a more inclusive and representative chaplaincy, and conversations about secular chaplains who represent a worldview more akin to that of most ADF personnel, suggest the future of Australian military chaplaincy may look quite different from its past.

Nevertheless, the Associated Protestant Churches, with more than a century's experience of providing unified chaplaincy within denominational diversity, is currently the one body that has the capability to ensure that the required number of chaplaincy positions are filled by suitable applicants. Time alone will tell how long it will be before other groups emerge to contribute new chaplains. It is, however, unlikely that the ADF will revert to a system based on proportionate representation, given the now well-accepted reality of military chaplaincy as generic rather than denominational. But in the meantime, the Associated Protestant Churches, successor to the original Other Protestant Denominations, once viewed as 'an aggregate of non-coherent and non-corporate factors having no solidarity or unified administration and responsibility', will continue to be the cornerstone that ensures the structure remains stable and functional.

---

11   Dexter, *Australia in the War of 1939–1945*, 354.

ABBREVIATIONS

AAChD	Australian Army Chaplains Department
AATTV	Australian Army Training Team Vietnam
ADB	Australian Dictionary of Biography
ADF	Australian Defence Force
ADFPCC	ADF Principal Chaplains Committee
AIF	Australian Imperial Force
ANZAC	Australian and New Zealand Army Corps
AOG	Assemblies of God
APCCC	Associated Protestant Churches Chaplaincy Board
ARA	Australian Regular Army
ARES	Army Reserve
AWAS	Australian Women's Army Service
AWM	Australian War Memorial
BCOF	British Commonwealth Occupation Force
BFBS	British and Foreign Bible Society
BOC	Board of Chaplains (RAAF)
CAA	Citizens Air Force
CFT	Continuous Fulltime Service
CIMHS	Critical Incident Mental Health Support
CMF	Citizens Military Force
CO	Commanding Officer
CPP	Commonwealth Parliamentary Papers
DCE	Dictionary of Christian Ethics
DCG	Deputy Chaplain General
DCO	Defence Community Organisation
DFCC	Defence Force Chaplaincy Committee
EATS	Empire Air Training Scheme
FUCCB	Federal United Churches Chaplaincy Boards
HMAS	Her Majesty's Australian Ship
HQ	Headquarters

HRA	Historical Records of Australia
HRNSW	Historical Records of New South Wales
LHQ	Land Headquarters
MBE	Member of the (Most Excellent Order of the) British Empire
MC	Military Cross
MEAO	Middle East Area of Operations
MID	Mentioned in Despatches
MOA	Memorandum of Arrangements
NAA	National Archives of Australia
NCO	Non-Commissioned Officer
NDT	New Dictionary of Theology
NSWGSAR	NSW Government State Archives & Records
OAM	Medal of the Order of Australia
OBE	Officer of the (Most Excellent Order of the) British Empire
OC	Officer Commanding
OPD	Other Protestant Denominations
PAC	Principal Air Chaplain
PCC-A	Principal Chaplains Committee-Army
PCC-AF	Principal Chaplains Committee-Air Force
PCC-N	Principal Chaplains Committee-Navy
PD	Protestant Denominations
POW	Prisoner of War
PRINCHAP-A	Principal Chaplain-Army
PRINCHAP-Air Force	Principal Chaplain-Air Force
PRINCHAP-Navy	Principal Chaplain-Navy
PTSD	Post Traumatic Stress Disorder
QMG	Quartermaster General
RAAF	Royal Australian Air Force
RAAChD	Royal Australian Army Chaplains Department
RACS	Religious Advisory Committee to the Services
RAN	Royal Australian Navy
RFD	Reserve Force Decoration
SDA	Seventh Day Adventist Church
UB	United Board
UC	United Churches
UCA	Uniting Church of Australia
VSUCCB	Victorian State United Churches Chaplaincy Board
WW1	World War One
WW2	World War Two
YMCA	Young Men's Christian Association

BIBLIOGRAPHY

A. MANUSCRIPT COLLECTIONS

1. Anglican Archives, Brisbane
Burgmann, E., Bishop of Goulburn. Letter to Archbishop Wand, 6 October 1939.

2. Baptist Union of New South Wales, Archives, Sydney.
Baptist Union of NSW Year Book, 1914–15.
Executive Committee Minutes, 15 April – 22 September 1915.

3. Baptist Union of Victoria, Archives, Melbourne.
Acton, Colin. 'Increasing Diversity Within the Navy Chaplain Branch: Respecting the Past, Creating the Future'. Unpublished paper.

Acton, Colin. 'Navy Chaplain Branch – Maritime Spiritual Wellbeing Officer (MSWO) Stream Position Paper'. Unpublished paper.

Associated Protestant Churches Chaplaincy Board Minutes, 9 March 2018 – 27 September 2019.

Federal United Churches Chaplaincy Board, *Minutes,* 25 November 1958 – 12 September 2017.

Victorian State United Churches Chaplaincy Board, *Minutes,* 21 December 1942 – June 1959.

Orr, Harry. Report on Visitation of Signal Line Morobe – Amboga Area.

Religious Advisory Committee to the Services, 'Comment on DGCHAP Navy OUT/2017, Increasing Diversity Within the Navy Chaplain Branch'.

'1989, 1990 Denominational Statistics and Chaplain Numbers,' in *United Churches Chaplaincy Historical Documents.*

4. Lutheran Church of Australia Archives, Adelaide

Stolz, Johannes. Address to UELCA.

5. Uniting Church of Australia Archives, Sydney.

Brookfield, J.W. 'President's Address,' Congregational Union of New South Wales Yearbook, 1969.

Congregational Union of NSW Yearbooks, 1916, 1940, 1945, 1956.

Lockley, Lindsay. 'Congregational Ministers in Australia, 1798–1977,' in Camden Theological Library.

Presbyterian Church of NSW, 'General Assembly Minutes,' June 1919.

B. OFFICIAL PUBLICATIONS

1. Australian Bureau of Statistics

ABS Data, 'Census of Population and Housing, 2011'.

2. Australian War Memorial

Chaplains [chronological by departure] (December 1914 – November 1918), AWM8 6/6/1.

Chaplains [chronological by departure] (December 1914 – December 1916), AWM8 6/6/2.

Chaplains [alphabetical], AWM8 6/6/3.

'AAChD Between Wars', in 'Military History of the Australian Army Chaplains', undated MS typescript, AWM 54, 177/2/1.

'AAChD – History 1939–46, United Churches,' in 'Military History of the Australian Army Chaplains', undated MS typescript AWM 54, 177/2/1.

Compulsory Military Service, AWM54, 946/2/1.

Papers of A. Treffry, AWM PR00032.

McIlveen, Arthur, AWM REL30722.002.

3. Diaries

Barwick, Archie, Canberra: AWM F940.26093 B296d.

McKenzie, William, Canberra: AWM 2019.22.2.

Miles, Frederick James, Senior Chaplain, Other Protestant Denominations, Headquarters AIF, London, August 1914 – October 1917, Canberra: AWM4, 6/4/1 Part 1.

4. Commonwealth of Australia.

Commonwealth of Australia Constitution Act, 1900, section 51(vi).

Commonwealth of Australia Defence Acts, 1903, 1927, 1941.

Commonwealth of Australia, Census, 3 April 1911.

Commonwealth Parliamentary Papers – General, 'Defence of Australia: Memorandum' Session 1910, Volume II.

Executive Minute 943, Commonwealth of Australia Gazette, 20 December 1913.

Murray, F.L. (ed.) *Official Records of Australian Contingents to The War In South Africa, 1899–1902*, (Department of Defence, Melbourne, 1911).

5. Department of Defence

Brooke, Allen. in *AAChD Post War Report*, Canberra: Army Office, 1945.

Department of Defence Media Release, 2 July 2013, 'Australia-led Combined Task Force concludes role with RAMSI'.

Memorandum of Arrangements Between the Chief of Defence Force Staff, Secretary, Department of Defence and the Heads of Churches Representatives, 1 July 1981.

Memorandum of Arrangements Between the Commonwealth of Australia and Heads of Churches Representatives, 2 December 2004.

Memorandum of Arrangements Between the Commonwealth of Australia and Heads of Churches Representatives, 2 December 2008.

Religious Advisory Committee to the Services Minutes, 23 September 1981 – 11 December 1990.

Royal Australian Air Force Board of Chaplains Minutes, 4 February 1965.

6. Historical Records of Australia

'Arrival of Lachlan Macquarie in Sydney, 28 December 1809,' in *Historical Records of Australia*, Series 1, Volume 7. Sydney: Library Committee of The Commonwealth Parliament, 1916.

George III, 'Commission to Arthur Phillip, 12 October 1786,' in *Historical Records of Australia*, Series 1, Volume 1. Sydney: Library Committee of the Commonwealth Parliament, 1914.

George III, 'Governor Macquarie's Instructions, 9 May 1809,' in *Historical Records of Australia*, Series 1, Volume 7. Sydney: Library Committee of the Commonwealth Parliament, 1916.

Macquarie, Lachlan. 'Despatch to Earl Bathurst, 4 December 1815,' in *Historical Records of Australia,* Series 1, Volume 8. Sydney: Library Committee of the Commonwealth Parliament, 1916.

Macquarie, Lachlan. 'Despatch to Earl Bathurst, 8 March 1816,' in *Historical Records of Australia,* Series 1, Volume 9. Sydney: Library Committee of the Commonwealth Parliament, 1917.

Macquarie, Lachlan. 'Despatch to Earl Bathurst, 18 March 1816,' in *Historical Records of Australia,* Series 1, Volume 9. Sydney: Library Committee of the Commonwealth Parliament, 1917.

Macquarie, Lachlan. 'Despatch to Earl Bathurst, 8 March 1816,' in *Historical Records of Australia,* Series 1, Volume 9. Sydney: Library Committee of the Commonwealth Parliament, 1917.

Macquarie, Lachlan. 'Despatch to Earl Bathurst, 18 March 1816,' in *Historical Records of Australia,* Series 1, Volume 9. Sydney: Library Committee of the Commonwealth Parliament, 1917.

Macquarie, Lachlan. 'Despatch to Earl Liverpool, 27 October 1810,' in *Historical Records of Australia,* Series 1, Volume 7. Sydney: Library Committee of the Commonwealth Parliament, 1916.

Paterson, William. 'Despatch to Rt Hon Henry Dundas, 21 March 1795,' in *Historical Records of Australia,* Series 1, Volume 1. Sydney: Library Committee of the Commonwealth Parliament, 1914.

Phillip, Arthur. 'Despatch to Lord Grenville, 7 November 1791,' in *Historical Records of Australia,* Series 1. Volume 1. Sydney: Library Committee of the Commonwealth Parliament, 1914.

Phillip, Arthur. 'Despatch to Under Secretary Nepean, 15 December 1791,' in *Historical Records of Australia,* Series 1, Volume 1. Sydney: Library Committee of the Commonwealth Parliament, 1914.

Phillip, Arthur. 'Despatch to Lord Grenville, 22 November 1791,' in *Historical Records of Australia,* Series 1. Volume 1. Sydney: Library Committee of the Commonwealth Parliament, 1914.

Phillip, Arthur. 'Despatch to Rt Hon Henry Dundas, 17 March 1792,' in *Historical Records of Australia,* Series 1, Volume 1. Sydney: Library Committee of the Commonwealth Parliament, 1914.

7. Historical Records of New South Wales

George III, 'Commission to First Chaplain, 24 October 1786,' in *Historical Records of New South Wales,* Volume 1, Part 2. Sydney: Charles Potter, Government Printer, 1892.

King, Philip. 'Despatch to Under Secretary Nepean, 25 October 1791,' in *Historical Records of New South Wales,* Volume 1, Part 2. Sydney: Charles Potter, Government Printer, 1892.

Phillip, Arthur. 'Despatch to Secretary Stephens, 12 May 1787,' in *Historical Records of New South Wales,* Volume 1, Part 2. Sydney: Charles Potter, Government Printer, 1892.

Phillip, Arthur. 'Despatch to Lord Sydney, 15 May 1788,' in *Historical Records of New South Wales,* Volume 1, Part 2. Sydney: Charles Potter, Government Printer, 1892.

Yonge, George. 'Despatch to Major Grose, 8 June 1789,' in *Historical Records of New South Wales,* Volume 1, Part 2. Sydney: Charles Potter, Government Printer, 1892.

8. National Archives of Australia

Air Board Agenda 767 (RAAF) – Conditions of service in the Chaplains Branch – Citizen Air Force, A14487, 7/AB/767.

Air Force Regulations to Provide for Appointment of 5 Staff Chaplains, A14487, 46/AB/6257.

Air Member for Personnel – Appointment of a permanent Chaplain policy, A705, 36/1/98.

Application for appointment as Staff Chaplains to RAAF – Rev T. C. Rentoul and A I Davidson and Rev Father Morrison and Rt Rev J. J. Brook, A705, 36/1/109.

Appointment of Anglican Replacements, A14487, 9/AB/1500.

Appointment of Chaplains, MP508/1, 56/750/340.

Appointment of Chaplains to Salvation Army and Other Denominations, A2023, A82/1/157.

Appointment of Lutheran Chaplains to Army, MP742/1, 56/1/17.

Appointment of OPD Chaplain General, MP508/1, 56/701/2.

Appointment of Roman Catholic Replacements, A14487, 9/AB/1478.

Appointment of Chaplains, Regulations for Chaplains, A705, 36/1/111.

Attachments Relating to Military Chaplains, B168, 1905/5614.

Australian Army Chaplains Department – Organisation Australian Army Chaplains Department – Organisation, MP508/1, 96/707945.

Ballard, H.R. Military Record, B883, SX22714.

Brooke, Allen. Military Record, B2455.

Chaplain General – Other Protestant Denominations, A2023, A82/1/191.

Chaplain RANR – Appointment of One Representing United Board, MP150/1, 431/202/187.

Chaplains Appointed to RAN and RANVR, MP150/1, 431/202/177 Part 1.

Chaplains – Conference Proceedings. Revision of Regulations Chaplains – Conference Proceedings. Revision of Regulations, A2023, A82/1/24.

Clarification of Chaplains or Welfare Workers (Salvation Army Personnel), MP508/1, 245/708/58.

Conference on appointment of Roman Catholic chaplains and Protestant chaplains to Royal Australian Navy in the Defence Forces, A2023, A82/1/2.

Cuttriss, G.P. Military Record, B2455.

Dempsey, J. Military Record, B2455.

Department of Army [Chaplains] – Senior Chaplain, United Board 3 M.D., MP508/1, 56/701/32.

Designation of the Commonwealth Naval Forces as Royal Australian Navy, MP1185/9, 559/201/574.

Fewster, F.M. Military Record, B883, WX37414.

Garland, A.W. Military Record, B883, VX32307.

Hansen, N.V. Military Record, B2455.

Helmore, R.A. Military Record, B883, VX197.

Henderson, K.T. Military Record, B2455.

John Curtin – Declaration of War on Japan, C102, 1552510.

Keyte, T.F. Military Record, B883, VX64453.

McKenzie, W. Military Record, B2455.

McIlveen, A.W. Military Record, B2455.

Methven, J.O. Military Record, B883, QX24324.

Miles, E.J. Military Record, B883, WX11128.

Miles, F.G. Military Record, B2455.

Organisation of Chaplains Dept. Military, B168, 1904/5368.

Orr, A.H. Military Record, B883, NX200382.

Perkins, H.S. Military Record, B2455.

Pickup, R.S. Military Record, B883, NX156703.

Procter, H.A. Military Record, B2455.

Regulations for Chaplains, A705, 36/1/111.

Religious Advisory Committee to the Services, A11114, 1981/49 PART 1.

Report of Chaplains General Stewart and Daws on their Visit to BCOF Japan, MP742/1, 56/1/99.

Ridley, J.G. Military Record, B883, NX167498.

Robertson, T.G. Military Record. B2455.

Salter, J.C. Military Record, B883, TX6008.

Senior Chaplain Salvation Army, A2023, A82/1/174.

Status and Advancement of Chaplains in the Time of War, A14487, 22/AB/3722.

Teece, A.H. Military Record. B2455.

Walden, G.T. Military Record. B2455.

Watts, C. Military Record, B883, SX8186.

Willings, H.J.W. Military Record, B884, N429336.

9. National Library of Australia

Downie, Graham, 'Interview with Bishop Morgan,' 24 July 1995, Audio.

Trathen, Douglas Arthur. Papers, 1939–1972, Canberra, NLA, MS 10546.

10. Sundry Sources Related to Government

Church Act 1837, *in Tegg's Pocket Almanac and Remembrancer – 1837,* Sydney: James Tegg, 1837.

New South Wales Government State Archives and Records. *On This Day, 7 Feb 1788 – Colony of NSW Formally Proclaimed.*

C. NEWSPAPERS AND PERIODICALS

1. Church

Australian Baptist, 12 May 1914 – 12 September 1939.

Australian Church Record, 2 July 1915.

Australian Christian, 27 August 1914 – 7 May 1966.

Australian Christian Commonwealth, 26 January 1940.

Catholic Church Press, 6 August 1914.

Church Record, 12 February 1941.

Methodist, 8 August 1914.

Presbyterian Messenger, 20 October 1916.

Victorian Baptist Witness, 5 December 1942.

War Cry, (Canada), 30 September 1899.

War Cry, (UK), 30 September 1899.

2. Newspapers

The Age, 5 August 1914 to 3 March 2015.

Argus, 29 April 1940.

Armidale Express and New England General Advertiser, 23 September 1940.

Army, 28 September 2000.

The Australian, 3 March 2015.

Bendigoan, 30 March 1916.

Brisbane Courier, 7 August 1914 – 26 April 1930.

Burrowa News, 17 April 1942.

Canberra Times, 9 October 1975.

News Digest, 28 July 1947.

Chronicle Adelaide, 4 November 1916 – 16 December 1916.

Courier-Mail, 24 March 1941.

Daily Advertiser, 13 March 1948.

Daily Telegraph, 10 August 1914 – 18 April 1940.

Diggers' Digest, 9 July 1948.

Freeman's Journal, 9 July 1948.

Herald-Sun, 3 March 2015.

Horsham Times, 12 February 1918.

The Mail, 17 April 1915.

Mercury, 14 May 1942.

Newcastle Morning Herald and Miners' Advocate, 24 May 1952.

Recorder, 18 October 1934.

Register, 2 March 1918 – 28 April 1930.

Scone Advocate, 30 April 1948.

Smith's Weekly, 3 December 1921 – 20 December 1925.

Solomon Times, 31 October 2007.

Sun (Sydney), 24 April 1972.

Sunday Times, 27 January 1929.
Sydney Morning Herald, 6 August 1914 – 17 May 1940.
Tribune, 10 October 1914.
Victorian Independent, August 1915 – September 1915.
Weekly Times, 26 June 1915 – 26 May 1917.
West Australian, 10 August 1914 – 17 April 1935.

3. Radio
Charles Bean, 'Fighting Mac,' *News Digest* broadcast, 28 July 1947.

D. CONTEMPORARY PRINTED MATERIALS
1. Books
2/4th Australian Infantry Battalion Association. *White over Green*. Sydney: Angus and Robertson, 1963.

Adam, P.J.H. 'Jesus,' in *New Dictionary of Christian Ethics and Pastoral Theology*, ed. David J. Atkinson and David H. Field. Leicester: IVP, 1995.

Baker, D. 'Lang, John Dunmore (1799–1878),' *Australian Dictionary of Biography*, Volume 2. Melbourne: MUP, 1967.

Bean, Charles. *Anzac to Amiens*. Canberra: Australian War Memorial, 1983.

Bettington, D.W. *Evangelicalism in Modern Britain: A History From The 1730s To The 1980s*. London: Routledge, 1989.

Blainey, Geoffrey. *A Short History of the 20th Century*. Melbourne: Viking, 2005.

Bolton, Barbara. *Booth's Drum: The Salvation Army in Australia, 1880–1980*. Sydney: Hodder and Stoughton, 1980.

Bond, John. *The Army that went with the Boys*. Melbourne: Salvation Army, 1919.

Bottrell, Arthur. 'Forbes, Arthur Edward (1881–1946),' *Australian Dictionary of Biography*, Volume 8. Melbourne: MUP, 1981.

Breward, Ian. *A History of the Australian Churches*. Sydney: Allen & Unwin, 1993.

Cable, K. J. 'Johnson, Richard (1753–1827),' *Australian Dictionary of Biography*, Volume 2. Melbourne: MUP, 1967.

Carlton, Mike. *Flagship*. Sydney: Random House, 2016.

Carlyon, Les. *Gallipoli*. Sydney: Macmillan, 2001.

Carlyon, Les. *The Great War*. Sydney: Macmillan, 2006.

Chapman, Graeme. *One Lord, One Faith, One Baptism; A History of Churches of Christ in Australia.* Melbourne: Vital, 1979.

Churchill, Winston. 'The Gathering Storm,' Vol 1, *The Second World War.* London: Casssell, 1948.Collins, David. *An Account of the English Colony in New South Wales,* Volume 1. Sydney: A.H. & A.W. Reed, 1975.

Clark, Manning. *A Short History of Australia.* Melbourne: Penguin, 2006.

------- *History of Australia,* abridged by Michael Cathcart. Melbourne: MUP, 1993.

------- *A History of Australia,* Volume 6. Melbourne: MUP, 1987.

Coulthard-Clark, C. 'Major-General Sir William Bridges: Australia's First Field Commander'. In *The Commanders, Australian Military Leadership in the Twentieth Century,* edited by D.M. Horner. Sydney: Allen & Unwin, 1984.

Colwell, J.E. 'Roman Catholic Theology'. *In New Dictionary of Theology*, edited by Sinclair B. Ferguson and David F. Wright. Leicester: IVP, 1988.

Cox, Harvey. *Fire from Heaven.* London: Cassell, 1996.

Crowley, F.K. *Modern Australia in Documents, 1901–1939.* Melbourne: Wren, 1973.

Cuttriss, George Percival. *Over the Top with the Australian 3rd Division.* London: Charles H. Kelly, 1918.

Davidson, Peter. *Sky Pilot, a History of Chaplaincy in the RAAF 1926–1990.* Canberra: Directorate of Departmental Publications, Department of Defence, 1990.

Dennis, Peter, Grey Jeffrey, Morris, Ewan, Prior, Robin, and Bou, Jean. 'Royal Australian Army Chaplains Department (RAAChD)'. In *Oxford Companion to Australian Military History.* Melbourne: OUP, 2008.

Devine, William. '*Story of a Battalion.* Melbourne: Melville and Mullen, 1919. Referenced in Johnstone, *The Cross of Anzac, Australian Catholic Service Chaplains.* Brisbane: Church Archivists' Press, 2003.

Dexter, David. *Australia in the War of 1939–1945: The New Guinea Offensives.* Canberra: Australian War Memorial, 1961.

Dunster, Nelson. *Padre to the 'Rats'.* London: Salvationist Publishing and Supplies,1971.

Edwards, John. *John Curtin's War,* Volume 1. Sydney: Viking, 2017.

Figley, Charles. *Stress Disorders Among Vietnam Veterans: Theory, Research and Treatment.* London: Brummer-Routledge, 1978.

Firkins, Peter. *The Australians in Nine Wars, From Waikato to Long Tan.* Sydney: Pan, 1982.

Fitzsimons, Peter. *Gallipoli*. Sydney: Random House, 2014.

Fitzsimons, Peter. *Tobruk*. Sydney: Harper Collins, 2006.

Fletcher, B.H. 'Phillip, Arthur (1738–1814),' *Australian Dictionary of Biography*, Volume 2. Melbourne: MUP, 1967.

Forbes, Cameron. *Hellfire, The Story of Australia, Japan and the Prisoners of War*. Sydney: Macmillan, 2005

Frame, Tom. *Losing My Religion–Unbelief in Australia*. Sydney: UNSWP, 2009.

Gammage, Bill. *The Broken Years, Australian Soldiers in the Great War*. Melbourne: Penguin, 1982.

Garrett, John and Farr, L.W. *Camden College: A Centenary History*. Sydney: Glebe, 1964.

Gladwin, Michael. *Captains of the Soul. A History of Australian Army Chaplains*. Sydney: Big Sky, 2013.

Grant, Lachlan. *Australian Soldiers in Asia-Pacific in World War 11*. Sydney: Newsouth Publishing, 2014,

Green, Vivian. *A New History of Christianity*. Stroud: Sutton Publishing, 1996.

Grey, Jeffrey. *A Military History of Australia*. Melbourne: Cambridge University Press, 2008.

Griffin, James. 'Mannix, Daniel (1864–1963),' *Australian Dictionary of Biography*, Volume 10. Melbourne: MUP, 1986.

Grudem, Wayne. *Systematic Theology, An Introduction to Biblical Doctrine*. London: Inter-Varsity Press, 2016.

Gunson, Niel. 'Albiston, Walter (1889–1965),' *Australian Dictionary of Biography*, Volume 7. Melbourne: MUP, 1979.

Ham, Paul. *Sandakan, The Untold Story of The Sandakan Death Marches*. Sydney: William Heinemann, 2012.

Ham, Paul. *Vietnam: The Australian War*. Sydney: Harper Collins, 2007.

Hastings, Max. *All Hell Let Loose, The World at War 1939-1945*. London: William Collins, 2012.

Heard, Barry. *Well Done Those Men*. Melbourne: Scribe, 2005.

Heiferman, Ronald. 'WorldWar II'. In *Wars of the 20th Century,* edited by S.L. Mayer. Secaucus, N.J: Derbibooks, 1975.

Holy Bible New International Version, Sydney: Bible Society in Australia, 2007.

Houison, A. *A Short History of St Philip's Church Sydney*. Sydney: St Philip's Vestry, 1910.

Hughes, Philip. *Charting the Faith of Australians: Thirty Years in the Christian Research Association*. Melbourne: Christian Research Association, 2016.

Hunt, Arnold. 'Howard, Henry (1859–1933)', *Australian Dictionary of Biography*, Volume 9. Melbourne: MUP, 1983.

Isaacs, Jeremy & Downing, Taylor. *Cold War*. London: Bantam, 1998.

Johnston, Mark. *Anzacs in the Middle East*. Melbourne: Cambridge University Press, 2013,

Johnstone, Tom. *The Cross of Anzac, Australian Catholic Service Chaplains*. Brisbane: Church Archivists' Press, 2003.

Judd, Stephen and Cable, K.J. *Sydney Anglicans*. Sydney: Anglican Information Office, 1987.

Keegan, John. 'Trench Warfare', in *History of the First World War*, Volume 2, edited by Barrie Pitt. London: Purnell, 1970.

King, H. 'Bourke, Sir Richard (1777–1855)', *Australian Dictionary of Biography*, Volume 1. Melbourne: MUP, 1966.

Lake, Marilyn and Reynolds, Henry. *What's Wrong with ANZAC?: The Militarisation of Australian History*. Sydney: New South, 2010.

Latourette, K.S. *History of the Expansion of Christianity*, Vol 7. London: Eyre & Spottiswoode, 1945.

Leske, Everard. *The Story of Lutherans and Lutheranism in Australia, 1838–1996*. Adelaide: Friends of Lutheran Archives, 2009.

Liddell Hart, B.H. *History of the Second World War*. London: Pan, 1973.

Linder, Robert, D. *The Long Tragedy: Australian Evangelical Christians and the Great War, 1914–1918*. Adelaide: Open Book, 2000.

MacDougal, A.K. *Anzacs Australians at War*. Sydney: Currawong, 1994.

Macquarie Dictionary, Sydney: Macquarie Library, 1981.

Macquarie, John. 'Just War'. In *A Dictionary of Christian Ethics*, edited by John Macquarrie. London: SCM, 1971.

McIntyre, Darryl. 'McIlveen, Sir Arthur William (1886–1979)', *Australian Dictionary of Biography*, Volume 15. Melbourne: MUP, 2000.

McKernan, Michael. *Australian Churches at War: Attitudes and Activities of the Major Churches 1914–1918*. Sydney: Catholic Theological Faculty and Australian War Memorial, 1980.

McKernan, Michael. 'McKenzie, William (1869–1947)', *Australian Dictionary of Biography*, Volume 10. Melbourne: MUP, 1986.

McKernan, Michael. *Padre*. Sydney: Allen & Unwin, 1986.

McKernan, Michael. *The Strength of a Nation*. Allen & Unwin, 2008.

McLachlan, N.D. 'Macquarie, Lachlan (1762–1824),' *Australian Dictionary of Biography*, Volume 2. Melbourne: MUP, 1967.

Moll, Hans. *Religion in Australia*. Melbourne: Thomas Nelson, 1971.

Neill, Stephen. *A History of Christian Missions*. Melbourne: Penguin, 1964.

Nutt, Dennis. *A Crucible of Faith and Learning: A History of the Australian College of Ministries*. Sydney: ACOM, 2017.

Overy, Richard. *World War 11, The Definitive Visual History*. London: Welbeck, 2020.

Parsons, Vivienne. 'Bain, James (1789-1794),' *Australian Dictionary of* Biography, Volume 1. Melbourne: MUP, 1966.

Pakenham, Thomas. *The Boer War*. London: Abacus, 1992.

Payne, Michael and Jessica Rae Barbera (eds.), *Dictionary of Cultural and Critical Theory*, 2nd edn. Chichester: Wiley-Blackwell, 2013.

Petras, Michael. 'Australian Baptists and The First World War in Retrospect,' in *Australian Baptists and World War I*, ed. Michael Petras. Sydney: Baptist Historical Society of N.S.W., 2009.

Piggin, Stuart and Linder, Robert. *Attending to the National Soul*. Melbourne: Monash University Publishing, 2020.

Pope, Steve. *Hornblower's Navy: Life at Sea in the Age of Nelson*. London: Orion, 1998.

Powell, Ruth, Sam Sterland and Miriam Pepper. *The Resilient Church: affiliation, attendance and size in Australia*. Sydney: NCLS Research, 2020.

Rennie, Ian S. 'Evangelical Theology,' in *New Dictionary of Theology*, edited by Sinclair B. Ferguson and David F. Wright. Leicester: IVP, 1988.

Reynaud, Daniel. 'Religion (Australia*),*' in *1914–1918: online. International Encyclopedia of the First World War*, eds. Daniel Ute, Peter Gatrell, Oliver Janz, et al. Berlin: Freie Universität, 2019.

Reynaud, Daniel. *Anzac Spirituality*. Melbourne: Australian Scholarly Publishing Pty Ltd, 2018.

Saunders, M. 'Rose, Herbert John (1857–1930),' *Australian Dictionary of Biography*, Volume 11. Melbourne: MUP, 1988.

Schedvin, C.B. 'Rivett, Albert (1885–1934),' *Australian Dictionary of Biography*, Volume 11. Melbourne: MUP, 1988.

Seifert, H. 'Peace and War'. In *A Dictionary of Christian Ethics*, edited by John Macquarrie. London: SCM, 1971.

Shakespeare, William. *Henry V,* Act 4, Scene 1.

Shermer, David. 'World War I'. In *Wars of the 20th Century,* edited by S.L. Mayer. Secaucus, N.J: Derbibooks, 1975.

Smith, N.C. *Home by Christmas.* Melbourne: Mostly Unsung, 1990.

Smyth, John. *In This Sign Conquer.* London: Mowbray, 1968.

Strong, Rowan. *Chaplains in the Royal Australian Navy, 1912 to the Vietnam War.* Sydney: UNSW Press, 2012.

Swain, Victor. *Australia: Moments in History.* Sydney: New Holland, 2011.

Tacey, David. *The Spirituality Revolution: The Emergence of Contemporary Spirituality.* Sydney: Harper Collins, 2004.

Taylor, A.J.P. 'The War in Perspective'. In *History of the First World War,* Volume 8, edited by Peter Young. London: Purnell, 1971.

'The First Away,' in *The RSL Book of World War 1*, edited by John Gatfield with Richard Landels. Sydney: Harper Collins, 2015.

The Treaty of London, 1839.

Thirkill, Angela. *Trooper to the Southern Cross.* Melbourne: Sun Books, 1966.

Thornton, Bruce. *And it Brought Forth Fruit: A History of the Association of Baptist Churches NSW & ACT.* Boston: Greenwood, 2020.

Vernon, P.V. *The Royal New South Wales Lancers:1885–1985.* Sydney: Royal New South Wales Lancers Centenary Committee, 1986.

Walker, Williston. *A History of the Christian Church.* Edinburgh: T.&T. Clark, 1959.

Wilkins, Arthur. *Life as I see It.* Brighton: Association for the Blind, 1988.

Woodland, Don. *Picking Up the Pieces.* Sydney: Macmillan, 2006.

Wright, D.F. 'Ecumenical Movement,' in *New Dictionary of Theology,* edited by Sinclair B. Ferguson and David F. Wright. Leicester: IVP, 1988.

Wurth, Bob. *The Battle for Australia.* Sydney: Pan Macmillan, 2013.

Yarwood, A. 'Marsden, Samuel (1765–1838),' *Australian Dictionary of Biography*, Volume 2. Melbourne: MUP, 1967.

Yuill, A. *Clerical Pioneers in New South Wales.* Sydney: Anglican Information Office, 1988.

2. Articles

Bottrell, Arthur. 'Australia's first two commissioned chaplains'. *Intercom,* No. 29 (December 1983).

Clayton, Elizabeth. 'Re-introducing Spirituality to Character Training in the Royal Australian Navy'. *Journal of the Australian Naval Institute* (2010).

Crotty, Martin. 'The Anzac Citizen: Towards a History of the RSL'. *Australian Journal of Politics and History,* (2007).

Cox, Lindsay. 'A tale of Two Armies'. *Halleluiah!* Volume 1, Issue 3 (Autumn 2008).

Dunn, Frederick. 'Christianity and Warfare'. *Australian Christian,* Volume 16 (1914).

Grant, Ivan. 'My Country, My Story: The Journey of an Aboriginal Army Chaplain'. *Australian Army Chaplaincy Journal*, No.27 (2016).

Gleeson, Gerald. 'A Christian Response to Torture,' unpublished paper, 2010.

Firth, John Rupert. 'Chairman's Address'. *Yearbook for 1940.* Sydney: Congregational Union of New South Wales, 1940.

Hawke, Bob. Address to the RSL Conference, August 1987.

Henderson, Kenneth Thorne. 'Memorandum issued for the personal use of C. of E. Chaplains serving in the A.I.F.'. Bendigo: Deputy Anglican Chaplain General.

Hoopman, Clemens. In *On Service with the Men and Women of the Evangelical Lutheran Church.* Adelaide: The Service Commission of the Evangelical Lutheran Church (n.d.).

Keyes, Don and Williams, Andrew. *Say a Prayer for Me: The Chaplains of the Vietnam War* [documentary film]. Sydney Headquarters Training Command Australian Army, 1995.

Kildea, J. 'What Price Loyalty? Australian Catholics in the First World War'. *Australasian Catholic Record* (2019).

Law-Davis, Harold. 'Padres Ponderings'. *Light Diet* (January 1945).

Main, Alexander. 'Thy God Reigneth'. *Australian Christian* 17.37 (17 September 1914).

------- 'Reconstruction and Theology'. *Australian Christian,* (1917).

Operation Gold, ADF Support to the Sydney 2000 Games. Canberra, Australian Defence Force, 2000.

Manley, Ken. 'Carry On! Victorian Baptists and World War Two'. *Our Yesterdays,* Volume 16 (2008).

Mislin, David. 'How Vietnam War Protests Accelerated the Rise of the Christian Right'. *Smithsonian Magazine* (3 May 2018).

Nutt, Dennis. 'A Life Well Lived'. *Historical Digest,* Issue 181 (November 2013).

------- 'Military Chaplains: For Service of our Soldiers'. *Australian Army Chaplaincy Journal,* No. 27 (2016).

O'Brien, G. 'The Empire's Titanic Struggle: Victorian Methodism and the Great War'. *Aldersgate Papers,* Volume 10 (September 2012).

Otzen, Ros. 'Major Trends Among Victorian Baptists 1939–65'. *Our Yesterdays,* Volume 16 (2008).

Porter, Patrick. 'The Sacred Service: Australian Military Chaplains in the Great War,' *War and Society,* Volume 20, (2002).

Reid, Stephen. 'Australia's Religious Communities: Numbers of people identifying with selected religious groups, 1911–2016'. Melbourne: *Christian Research Association* (n.d.).

Sabel, E.T. 'A History of Character Guidance in the Australian Army,' *Australian Defence Force Journal,* No.28 (1981).

Smith, Robert. 'Report on the International Military Chiefs of Chaplains Conference Prague 31 January – 4 February 2011,' *RACS, Minutes,* March 2011.

Vader, John. 'The Anzacs'. *History of the First World War,* Volume 3, ed. Barrie Pitt. London: Purnell (1970).

White, Hayden. "The Question of Narrative in Contemporary Historical Theory" in *History and Theory,* Vol.23, No.1 (Feb. 1984): 1-33.

Winkler, M.H. 'A Padre Speaks'. *On Service with the Men and Women of the Evangelical Lutheran Church.* Adelaide: The Service Commission of the Evangelical Lutheran Church of Australia (n.d.).

Woodbury, David. 'Do You Think I'm Afraid to Die with You'. *Halleluiah,* Volume 1, Issue 3 (Autumn, 2008). 'When the World Went to War, the Salvation Army Was There'. *Halleluiah,* Volume 1, Issue 3 (Autumn, 2008).

Yesberg, Barrie. 'Towards Spiritual Health: The Australian Defence Force Spiritual Health and Wellbeing Strategy'. Unpublished paper, 15 September 2014.

E. THESES

Abbott, Douglas. 'In This Sign Conquer: The Chaplains General of the Australian Army, 1913–1981'. Unpublished manuscript, 1995.

Burley, Roy. 'The Age of Negligence? British Army Chaplaincy 1796–1844'. Master thesis, University of Birmingham, 2013.

de Reland, Elizabeth. 'Holiness and Hard Work: A History of Parramatta Mission, 1815–2015'. PhD thesis, Charles Sturt University, 2018.

Jackson, Hugh Rutherford. 'Aspects of Congregationalism in South-Eastern Australia, circa 1880–1930'. PhD thesis, ANU, 1978.

Tippett, R.W. 'Australian Army Chaplains, South West Pacific Area, 1942–1945'. Master thesis, University of New South Wales, 1989.

F. INTERVIEWS AND CORRESPONDENCE

1. Oral History

Holmes, Denby	16 April 2021, 30 October 2022; 14 November 2022.
McNamara, Philip	1 November 2022.
Norris, Harold	October 1986
Ogden, Ralph	10 September 1986
Pither, Keith	24 August 1986
Prior, Alan	25 July 1986
Starr, Frank	25 July 1986
Willis, Mark	16 February 2023.

Taped interviews with Rodney Tippett, in Tippett, *Australian Army Chaplains, South West Pacific Area, 1942–1945*.

2. Correspondence

Air Vice Marshall Parker to Chaplain General McCulloch, 26 January 1978.

Assemblies of God in Australia to Federal United Churches Chaplaincy Board, 27 June 1988.

Assistant Minister for Defence to Robert Smith, 10 December 2014.

Australian Pentecostal Ministers Fellowship to Federal United Churches Chaplaincy Board, 6 June 1991.

Allan Brook to Federal United Churches Chaplaincy Board, 23 July 1947.

Baptist Union of Australia to Federal United Churches Chaplaincy Board, 31 May 1992.

Chief of Defence to Australian National Imams Council, 11 March 2015.

Congregational Union of Australia to Federal United Churches Chaplaincy Board, 12 June 1963.

Craig Willmott to Robert Smith, 9 October 2022.

Don Woodland to Robert Smith, 26 May 2021.

E. Seaton to Rodney Tippett, 17 January 1987, in Tippett, *Australian Army Chaplains, South West Pacific Area, 1942–1945*.

Eric Hollard to Rodney Tippett, 1 April 1986, in Tippett, *Australian Army Chaplains, South West Pacific Area, 1942–1945*.

Ernest Sabel to Robert Smith, 7 August 2021, 17 November 2022.

Estherby, Ralph to Robert Smith, 4 August 2022.

G. Leister to Rodney Tippett, 26 February 1987, in Tippet, *Australian Army Chaplains, South West Pacific Area, 1942–1945*.

Geoff Crossman to Douglas Abbott, 10 April 1989, in Abbott, *In This Sign Conquer: The Chaplains General of the Australian Army, 1913-1981*.

Griffiths, David to R.L. Coombe, 2 June 1988.

H.R. Ballard to Douglas Abbott, 14 August 1986, in Abbott, *In This Sign Conquer: The Chaplains General of the Australian Army, 1913–1981*.

J. Tan to Robert Smith, 8 June 2021.

Jaensch, Darren to Robert Smith, 1 December 2022.

Jason Wright to Robert Smith, 20 March 2023.

Keith Pither to Rodney Tippett, 24 August 1986, in Tippett, *Australian Army Chaplains, South West Pacific Area, 1942-1945*.

Ken Jarvis to the Federal United Churches Chaplaincy Board, 3 November 1975.

Ken Jarvis to David Griffiths, 24 May 1976.

McCullough to his wife, in Tippett, *Australian Army* Malcolm *Chaplains, South West Pacific Area, 1942-1945*.

Mark Willis to Robert Smith, 9 February 2023.

Nigel Long to Robert Smith, 28 March 2022.

Naval Board to Federal United Churches Chaplaincy Board, 15 January 1974, 18 April 1974 and 10 June 1974.

North Queensland Area Chaplains' Committee to Religious Advisory Committee to the Services, 9 April 1990.

R.L. Coombe to Geoff Crossman, 11 November 1987.

Ralph Ogden to Rodney Tippett, 10 September 1986, in Tippett, *Australian Army Chaplains, South West Pacific Area, 1942–1945*.

Russell Mutzelburg to Robert Smith, 8 February 2023.

Salvation Army Australian Southern Territory to Federal United Churches Chaplaincy Board, 29 March 1987.

Salvation Army Southern Territory to Federal United Churches Chaplaincy Board, 21 May 1992.

INDEX

A

Afghanistan War | 4, 204-205, 208.
Affiliated Representatives Committee | 217.
Albiston, Principal Chaplain Walter | 116, 123-24, 141, 144-45, 149, 151, 161, 163-64, 167, 174, 181, 191, 198.
Alcock, Chaplain Michael | 195, 215.
Alexander, Chaplain Ray | 173.
Ashworth, Chaplain George | 176.
Assemblies of God/Australian Christian Churches | 121, 194-95, 213, 215, 224.
Associated Protestant Churches Chaplaincy Board (APCCB) | 223-24.
Australian Army Chaplains Department | 22-3, 216, 225.
Australian Pentecostal Ministers Fellowship | 194-95.
Australian Regular Army | 160, 170, 205, 228.

B

Bain, Rev James | 8-10.
Baker, Chaplain Melissa | 207.
Ballard, Chaplain Hugh | 147, 167.
Baptist Union | 32-3, 101, 118, 139-40, 164, 195, 215.
Barrie, Admiral Chris | 204.
Bedford, Chaplain Roy | 185.
Birdwood, General Sir William | 66, 93.
Blamey, General Sir Thomas | 122.
Bosch, Chaplain Kees | 216.
Boer War | 16, 26, 35, 48, 50.
British Commonwealth Occupation Force | 170.
Brooke, Chaplain General Allen | 119-23, 125, 127, 131-33, 137, 141, 144, 146, 149, 160-67, 171, 174-76, 181, 198, 201.
Brookfield, Rev JW | 187-88.
Brown, Chaplain Glen | 177, 181.

C

Cartwright, Rev Robert | 8.
Chaplains Branch RAAF | 124.
Chaplains General, Establishment of | 19.
Chauvel, Lieutenant General Sir Harry | 22, 60.
Churches of Christ | 23, 33-4, 38, 40, 43-46, 49-50, 66, 71, 73-4, 94, 98, 111, 115-16, 121, 127, 138-39, 164, 167, 179, 188, 197, 224, 227.
Clark, Chaplain Harry | 161.
Cold War | 4, 159, 177.
Confrontation with Indonesia | 176.
Congregational Church | 14-5, 23, 34-5, 45, 60, 67, 73, 91-2, 98, 100, 110, 124, 138, 167, 187-88, 193, 224, 226.

Conolly, Fr Philip | 12.
Collingridge, Fr Charles | 15.
CO's Hours and Character Guidance Courses | 141, 154, 173-76.
Cowper, Rev William | 8-10.
Crossman, Principal Chaplain Geoff | 120, 168, 175, 181, 191-92, 198, 200-202, 229.
Crossman, Chaplain William | 120, 138-39.
Crowley, Doctor Desmond | 188.
Cuttriss, Chaplain George | 50, 65, 72, 81, 95.

D

Davies, Rev Rowland | 14.
Davies, Chaplain E | 79.
Defence Force Chaplains College | 196-97.
Demobilization | 123, 163.
Dempsey, Chaplain J | 51.
Dillon, Chaplain Howard | 185.
Doecke, Chaplain Gary | 205.
Dunmore Lang, Rev John | 12.

E

Fahey, Chaplain John | 57.
Farr, Chaplain Alan | 170.
Federal United Churches Chaplaincy Board (FUCCB) | 108, 123-24, 126-28, 130, 135-36, 140-41, 146, 149-50, 152-54, 160, 163-64, 167-69, 171, 173, 178, 186-87, 190-96, 198-202, 206-208, 210, 212-16, 222-24, 227-28.
Fewster, Chaplain F | 147.
Forward Defence Policy | 160, 170.

G

Gallipoli Campaign | 33, 59, 61, 71-2, 85.
Garland, Corporal Alan | 149.

Genende, Rabbi Ralph | 217.
Gillison, Chap Andrew | 58.
Gleeson, Fr Gerry | 209.
Gomm, Chaplain Leslie | 170-71.
Gough, Archbishop Hugh | 187.
Grant, Chaplain Ivan | 207.
Gray, Albert | 173.
Griffiths, Principal Chaplain David | 192, 195, 198-99, 215.
Grulke, Chaplain David | 205.

H

Haig, Filed Marshall Sir Douglas | 65.
Hansen, Chaplain General Norman | 118-19, 121, 123, 125, 166.
Heard, Barry | 180.
Helmore, Chaplain R | 132, 138-39, 145.
Henderson, Chaplain K | 131.
Holmes, Chaplain Denby | 176-77, 179, 181, 184-85.

I

Inches-Ogden, Principal Chaplain Catie | 207.
Iraq War | 204, 206, 208. 230.

J

Jarvis, Chaplain Kenneth | 175-77, 190-92, 197-99, 201-202, 229.
Johnson, Rev Richard | 4, 6-7, 10-11, 26.
Just War Theory | 31, 34, 155, 187.

K

Keyte, Chaplain T | 147.
Knopwood, Rev Robert | 17.
Knox, Reverend Broughton | 187.
Korean War | 171-74.

L

Leahy, Lieutenant General Peter | 205.
Leigh. Rev Samuel | 11.
Liddell, Chaplain Edmond | 177.

Lock, Principal Chaplain Garry | 210, 223-24, 229.
Loy, Dr Allan | 187.
Lutheran Church | 24, 108, 115, 124-26, 153, 164-65, 181, 188, 224.

M

Malayan Emergency | 160, 176.
Mack, Chaplain David | 176.
McAdam, Chaplain George | 173-74.
McCullough, Chaplain General Malcolm | 144, 155, 166-67, 169, 178, 181, 190, 193-94, 198, 201.
McIlveen, Chaplain Arthur | 132-135.
McKenzie, Chaplain Raymond | 169.
McKenzie, Chaplain William | 43, 48, 50-4, 56-9, 62-4, 68, 71-2, 75-6, 80, 84, 88, 90, 93-4, 133.
McNamara, Brigadier Philip | 179.
Macquarie, Governor Lachlan | 9-12.
Marsden, Rev Samuel | 7, 11.
Memorandum of Arrangements between Defence and the Churches | 189, 194-95, 215.
Methven, Chaplain E | 147.
Middle East Campaign | 113, 120, 132-36, 153, 204.
Miles, Chaplain E | 132.
Miles, Chaplain Frederick | 21, 42, 45, 48, 50-6, 59, 61-4, 67-72, 75, 79-81, 87-9, 92-3, 100.
Mitchell, Chaplain Mairi | 207.
Monash, Lieutenant General Sir John | 66, 81.
Moral Leadership | 172, 175-76.
Morgan, Chaplain General JA | 162.
Mutzelburg, Principal Chaplain Russell | 205, 223, 229.

N

Nicholson, Chaplain John | 173.

O

Orr, Chaplain Harry | 147-48.
O'Flynn, Fr Jeremiah | 12.
Other Protestant Denominations/OPD | 3-4, 13-4, 21, 24, 26, 32, 37, 39-42, 44-6, 50, 65-7, 69-70, 72, 75, 78, 80-2, 84, 86, 88-90, 92, 94-5, 98, 117, 126, 130, 169, 196, 211, 225, 228, 231-32, 236-37.

P

Paterson, Andrew (*Banjo*) | 53.
Pax Christi | 187.
Peace Keeping Operations | 203.
Perkins, Chaplain Harold | 67, 79, 92. 99.
Perkins, Chaplain Thomas | 49.
Phillip, Governor Arthur | 5-10,17.
Pickup, Chaplain R | 147.
Principal Chaplains | 20, 23, 189-92, 199-201, 214, 223, 229.
Procter, Chaplain Henry | 52, 78, 95.
Proportionate Representation | 102-103, 114, 122, 160. 165. 210, 214-15, 227-28, 232.
Protestant Ministry to the ADF | 214.

R

Reidel, Chaplain Erich | 165.
Religious Advisory Committee to the Services (RACS) | 121, 166-67. 189, 194-96, 198-201, 208-10, 212-17, 219-20.
Richardson, Chaplain Kenneth | 176.
Ridley, Chaplain John | 137, 141.
Riley, Chaplain General Charles | 43-4, 89, 130-31, 165, 179, 227.
Riley, Chaplain Douglas | 151.
Robertson, Edwin | 173.
Robertson, Chaplain Theodore | 51, 58, 99.

Rose, Chaplain Herbert | 15.
Rothwell, Chaplain Arthur | 175.
Roy, Chaplain Bruce | 199.

S

Sabel, Chaplain Ernest | 177, 185, 192, 197-202.
Salter, Chaplain J | 132-33.
Salvation Army | 23, 35-6, 40-2, 45-6, 48, 50, 52, 66, 73, 76, 93-4, 98, 100, 115-16, 127, 132-33, 135-36, 139, 145, 148, 154, 164, 173, 178, 186, 188, 195, 224, 231-32.
Sanderson, Lieutenant General John | 203.
Segelman, Chaplain Yossi | 220.
Seventh Day Adventist Church | 194, 224.
Sharteris, Chaplain Martin | 193.
Sinai Campaign | 51, 60.
Smith, Chaplain Robert | 204, 210, 215, 219, 223.
Snook, Chaplain Wendy | 207.
Southwest Pacific Campaign | 113, 136, 143, 153-54.
Stevenson, Chaplain General Alexander | 88, 161-62.
Storie, Rev John | 14.

T

Tange, Sir Arthur | 189, 191.
Teece, Chaplain Ashley | 50-2, 54, 60-1, 64, 67, 70, 91-2.
Therry, Fr John | 12.
Thompson, Chaplain Lester | 184.
Tippett, Chaplain Rodney | 132, 135, 142, 176.
Tobruk | 120, 133-34.
Trathen, Reverend Douglas | 187.
Treffry, Andrew | 180.

U

United Board | 98-105; 108; 110; 112-24; 127; 130; 133; 135-36; 138; 150-54; 227-28.

V

Vale, Rev Benjamin | 10.
Vietnam War | 160, 177-80, 184, 186-88, 202, 204, 206-207.

W

Walden, Chaplain George | 21, 45, 48, 51-2, 54, 58-9, 61, 63-4, 69, 71, 76-7, 87, 94, 99.
Walker, Reverend Alan | 187.
Watts, Chaplain Charlie | 132.
Western Front | 39, 42, 51, 61, 72, 77, 90.
Whitley, Chaplain Ian | 205.
Wilkins, Chaplain Arthur | 152, 168, 191.
Willings, Chaplain Horace | 149.
Willis, Principal Chaplain Mark | 205, 223, 229.
Willis, Chaplain Peter | 223.
Willmott, Chaplain Craig | 205.
Woodland, Chaplain Donald | 178-9, 186.

Y

Yesberg, Principal Chaplain Barrie | 205, 209, 229-30.

www.ingramcontent.com/pod-product-compliance
Lightning Source LLC
Chambersburg PA
CBHW041136110526
44590CB00027B/4041